Textile and Laundry in Hotel Industry

Textile and Laundry in Hotel Industry

Naresh Grover

RANDOM PUBLICATIONS
NEW DELHI (INDIA)

Textile and Laundry in Hotel Industry

ISBN 978-93-5111-871-8

Published in 2016 in India by

RANDOM PUBLICATIONS

4376-A/4B, Gali Murari Lal, Ansari Road
New Delhi-110 002
Phone : +9111-43580356, 011-23289044, 011-43142548
e-mail: sales@randompublications.com,
info@randompublications.com, randomexports@gmail.com

Reprinted 2023

Type Setting by : Friends Media, Delhi-110089
Digitally Printed at : Replika Press Pvt. Ltd.

Preface

Laundering is one of the most important part of housekeeping operations. Almost every large hotel has in-house laundry facility. In the 'good old days' laundry was done with soaps, scrub boards, wash tubs and lots of elbow grease. Today's advanced washing machines, hi-tech detergents and laundry additives make wash days easier and more successful, especially with the array of fabrics used in contemporary clothing.

Laundry was probably first done by immersing cloth in streams and letting the stream carry away the materials causing stains and smells. Laundry may still be done this way in some less industrialized areas and rural regions.

Agitation helps remove the dirt, so the laundry is often rubbed, twisted, or slapped against flat rocks. Various chemicals may be used increase the solvent power of water, such as the compounds in soaproot or yucca-root used by Native American tribes. Soap, a compound made from lye (from wood-ash) and fat, is an ancient and very common laundry aid. However, modern washing machines typically use powdered or liquid laundry detergent in place of soap.

Textiles have an assortment of uses, the most common of which are for clothing and containers such as bags and baskets. In the household, they are used in carpeting, upholstered furnishings, window shades, towels, covering for tables, beds, and other flat surfaces, and in art.

The practical function of clothing is to protect the human body from dangers in the environment: weather (strong sunlight, extreme heat or cold, and precipitation, for example), insects, noxious chemicals, weapons, and contact with abrasive substances, and other hazards. Clothing can protect against many things that might injure the naked human body. In some cases clothing protects the environment from the clothing wearer as well (example: medical scrubs).

This book is an indispensable guide to caring for all cloth in a hotel: from kitchen rags to bedding, hand-washables, and table clothes to window curtains. It offers detailed guidance on selecting and maintaining hotel clothing and fabrics, installing carpets, curtains and rugs, arranging table linens, removing stains, and making environmentally informed choices.

– Author

Contents

1

Introduction

The maintenance of a textile product after purchase is of prime interest to the consumer and to commercial fabric care operations. The major factor to be considered is cleaning and soil removal of the textile during continued use. In order to have a fuller understanding of the cleaning process, one must examine the nature of textile soils, detergency, and soil removal, and the wet or dry cleaning processes used.

USES OF TEXTILES

Textiles have an assortment of uses, the most common of which are for clothing and containers such as bags and baskets. In the household, they are used in carpeting, upholstered furnishings, window shades, towels, covering for tables, beds, and other flat surfaces, and in art. In the workplace, they are used in industrial and scientific processes such as filtering. Miscellaneous uses include flags, backpacks, tents, nets, cleaning devices such as handkerchiefs and rags, transportation devices such as balloons, kites, sails, and parachutes, in addition to strengthening in composite materials such as fibreglass and industrial geotextiles. Children can learn using textiles to make collages, sew, quilt, and make toys.

Textiles used for industrial purposes, and chosen for characteristics other than their appearance, are commonly referred to as *technical textiles.* Technical textiles include textile structures for automotive applications, medical textiles, geotextiles, agrotextiles, protective clothing (*e.g.* against heat and radiation for fire fighter clothing, against molten metals for welders, stab protection, and bullet proof vests). In all these applications stringent performance requirements must be met. Woven of threads coated with zinc oxide nanowires, laboratory fabric has been shown capable of "self-powering nanosystems" using vibrations created by everyday actions like wind or body movements.

Once a textile fabric has been manufactured, its potential for causing damage to the environment does not end. A fabric, no matter how high its quality or how well it is finished, coloured and generally enhanced, is of little or no use unless it is made into something. The possible products of this further processing include such widespread objects as clothing, upholstery, drapery, bedding and industrial goods. We normally tend to regard the primary (*i.e.* fabric production) and secondary (*i.e.* products made from fabrics) textile industries as separate ones, but the environment makes no such distinction; pollution is pollution no matter what its source may be.

One of the major problems facing the entire industry today is a lack of communication between these two different areas, since there is often a misinterpretation of the needs of the secondary sector by members of the first one, as well as a misunderstanding of the capabilities of the primary industry by the secondary one. Nature may benefit substantially by the development of a closer connection between the two, reducing the number of discarded materials that result from these failures to match needs, abilities and production.

TYPES OF USE

A distinction is proposed here between different types of use in a way not normally adopted. Those uses intended to protect the environment in some way will be separated from the others and treated. This is obviously not a simple matter, because there will be immense grey areas where the two cannot readily be distinguished. A geotextile barrier designed to prevent flooding of houses in an area inundated with water from a burst river bank, for instance, is clearly intended to keep the human inhabitants of the threatened region safe and alive. A geotextile used to contain polluted substances in a landfill site from leaching out into the water table is intended to protect the environment.

Yet, the flood barrier also prevents erosion of land by water flow, while the contained pollution is also prevented from poisoning human beings, so that both end results are met in each case. The distinction here (which is totally arbitrary and personal) is to determine what seems to be the more important criterion for use and to allocate the application to the section that makes most sense in that case.

NORMAL USES

Clothing

The most obvious normal use of textiles is in making clothing, a use that has been around for thousands of years. Products of the garment sector will include only clothing and accessories. As always, the machines needed to make them are complex, though usually much smaller in scale than those used in textile yarn and fabric manufacture. Cutting, sewing, pressing and other machines all need to be made and operated, at the now-familiar cost to the planet. There are frequently excess pieces of fabric that are cut away in making clothing to be discarded. It is, admittedly, possible to recycle these scraps, but large, complex machinery is needed to put them back into the production train. Although such scraps may be too small anyway to make recycling worthwhile, one author describes yarns made with fibres recovered from the garment industry, noting that they avoid excess pollution by eliminating the need for fertilisers and dyes. Uses as socks, blankets and upholstery materials are mentioned specifically, a point that gives the key to this over-optimistic attitude. These articles differ from most textile end uses in that colour is not critical, unlike virtually all other applications. In more usual recycled goods, especially as they are almost invariably coloured, some process for removing all the dyestuffs is always needed. This involves chemical reactions in which bleaching agents or dye strippers are used to destroy the dye molecules or change their molecular structure to render them colourless. These substances are waste products that can harm the Earth once discarded into the environment.

The Fashion Industry

One of the most powerful driving forces influencing the manufacture of clothing is the fashion industry, which has increased considerably in importance in the centuries since the Industrial Revolution.Clothing is one of the easiest ways to display wealth conspicuously, so it is now accepted among people in many parts of the world (and particularly in the developed nations) that new outfits should be purchased at every season of the year.

Seasonal variations in climate are an added major incentive for these expenditures and, in practice, the resulting annual production of clothing may be regarded as too great to be justified as environmentally responsible. Few people actually wear more than one outfit at a time, which means that many garments are stored in environmentally destructive plastic bags for a great proportion of their useful life.

After the season ends, it is currently fashionable, in a kind of lip service to planetary well-being, to 'recycle' the articles via alternative outlets, such as second-hand clothing stores or factory shops selling designer clothes at reduced prices. Even this gesture is not entirely altruistic in terms of the Earth's welfare; the loss of an article of clothing invariably leaves a gap that must be filled by something new in the wardrobe of the person making this supreme sacrifice. The premises where the unwanted garments are resold also need to be built, heated, lit, staffed and so on. Unfortunately, the pressures of the fashion industry appear to be increasing rather than declining, as would be preferable for ecological welfare. The problem, as always, lies in commercial interest. Modern people, especially in the developed world, are accustomed to the need to follow a fashion cycle on an annual or even seasonal basis, so that the fashion industry (and those connected with it) has a vested interest in ensuring a regular turnover of garments.

It is inevitable that a large reduction in sales would lead to a major loss in employment, because it is not only the textile and clothing manufacturers who would be out of work, but also the designers, alteration staff, marketing specialists, transportation personnel, retailers, magazine editors or artists, and anyone else with an ancillary interest in the fashion world. The cumulative psychological effect on Society of the need of all these people to be kept in employment is difficult to resist, meaning that the relentless drive to be up-to-date will continue for as long as we (or our planetary home) are prepared to tolerate it.

Environmental Aspects

The renewal of the fashion requirement per se does not merely cause environmental problems in the debate regarding recycling versus disposal. As a consequence of our modern lifestyle and technology, dust and dirt are omnipresent, and soiling necessitates emergency procedures in the form of laundering or dry cleaning.

Both of these can cause environmental difficulties. Laundering involves exposing the textile article to the combined effects of water, heat, agitation and detergent. Often, an optical bleach or a fabric softener is added to the mix, and machine drying is frequently used. Both of these procedures give a much more luxurious feel than do an untreated wash and outdoor drying on a clothing line.

The energy needed to heat the water and operate the machinery, together with the chemicals discarded into the drain, are a source of harm, as also are the heat needed to operate and the polluted exhaust emitted from the dryer. The thermal load from both washing and drying can also affect the surrounding air or water, making it unsafe for other species to exist in the neighbourhood of the discard point unless precautions are taken to reduce the rise in temperature occurring on discharge.

Dry Cleaning

Dry cleaning also has ecological consequences. The machinery needs to be built and operated, while the solvents used for extracting the dirt are often toxic or carcinogenic. Even though strenuous efforts are made to contain them and recycle them for later use, there is still an inevitable loss because of entrapment in the fibres or escape into the air via the garment transport path. Jipp describes nontoxic stain removal using solvents without chlorocarbons, so that the chemicals can be regarded as being kind to both fibre and environment. Nevertheless, all of the reagents used in maintenance, even these more harmless ones, have to be manufactured, with expensive loading on the environment.

Maintenance Chemicals

The actual chemical substances used in maintenance can be examined in more detail to ascertain their potential for harm. As a consequence of our current fascination for cleanliness, garments are washed or cleaned far more often than is necessary for the sake of health or even hygiene. For this reason, the quantities of detergent, softeners, bleaching agents, dry cleaning solvents or other chemicals expelled are enormous.

Detergents contain alkalis and organic chemicals that act as pollutants. They also often contain phosphates, used as 'builders' to enhance the effectiveness of the washing action, which are known to encourage the growth of algae in large bodies of water.

These can take over the oxygen available in the water, preventing other species (both animal and vegetable) from having access to it, As a result, fish can die and the balance of the local aquatic plants can be disturbed seriously, bringing about major changes in the locality and eventually resulting in a dead body of water where fish no longer exist and weeds choke the entire surface. Fabric softeners, usually quaternary ammonium compounds, are becoming more and more popular. These bring about water pollution. They are produced by relatively complex chemical reactions, with the usual environmental concerns and their disposal brings about harmful changes in the local water supply.

Household Textiles

In the home, textiles find widespread use, from household furnishings such as carpets, cushions and table linens, to towels, bath mats or dish cloths. David Rigby Associates give an overview of the various types of textile applications in home furnishing products, suggesting that the largest growth in the foreseeable future will occur in non-wovens or fibrefill categories, rather than in the more traditional woven or knitted fabrics. Two unusual applications in the household category have been reported in the literature; one author writes of the usefulness of wool tightly woven and attached to gypsum board (in the form of decorative panels) as a means of combating the relatively new hazard of sick building syndrome by absorbing the formaldehyde reputedly responsible for this problem. The second application consists of a system used to wrap houses during construction to maintain acceptable working comfort, for craftsmen operating on the outside walls of the house in adverse weather conditions, with enough moisture vapour permeability to ensure that breathing comfort can be retained.

Technical Textiles

Bhonde gives an overview of textiles used in applications beyond those of clothing or household purposes, while Legler examines technical textiles from the perspective of recent advances. Both authors more or less agree on the list of end uses, including geotextiles, agrotextiles, industrial products (including composite reinforcement), automotive, space, protective and medical ones. Not mentioned specifically in this list is the use of textiles in building, which may be extended to the subject of architectural fabrics. Their adoption, according to Hill, reduces building time significantly and saves natural resources (a valuable advantage from our point of view!), while providing long life and cost-effective, aesthetically pleasing appearance.

Swedberg describes applications of tent structures using PTFE-coated fibreglass with steel tubing and cables and, in a later paper, the use of air-inflated beams for supporting massive structures in high wind or loading conditions with potential applications in aerospace or nautical projects. Other interesting applications are the use of a composite of high-tensile glass with epoxy resin to make a reinforcement beam as protection against earthquake collapse and wrapping corroded bridge columns with carbon fibre to prevent further damage.

House Construction

House construction can also be aided indirectly by fibres, when used as reinforcement for concrete or in other aspects of building. In such applications, the environment is preserved by preventing premature failure of the material, allowing it to last longer. Peter recommends Kevlar for concrete reinforcement, because it imparts low density, non-catastrophic failure (and resistance to repeated impacts) to the concrete. In addition, because the fibres are electrically insulating, Kevlar permits building to take place more safely near high-voltage power lines.

Komatsu assess the durability of geotextiles in reinforcing asphalt concrete, noting that this process increases viscosity of the material and reduces the stress concentration of wheel loading. Locatelli reports the production of a stretchable roofing felt that can be installed and maintained without shrinkage when used in a layered system. A French company is marketing a woven structure that can be laid across soft, sandy or swampy ground, allowing vehicles to pass safely. In temporary houses, such as tents, fabrics are invariably used for their combination of light weight, easy folding ability, weather (especially wind) resistance and versatility.

Motor Vehicles

Fung provides a detailed description of the many applications of textiles, in the form of coated or laminated fabrics, in vehicles. He includes in his listing seat materials, headliner structures, other interior coverings, air bag components, convertible coverings, liners for bonnets or wheel arches, carpeting, noise control items and drive belts. In view of current traffic hazard concerns involving motor vehicles, the air bags used to enhance safety in crashes have received particular attention. One writer recommends the use of nylon 4.6 in making them, because of its better thermal stability.

Barnes and Rawson, after devising a new test to assess the efficiency of coated fabrics for this end-use, find that nylon 6.6 is better than either nylon 6 or polyester in this regard. They also report that a silicone-based coating should be applied to the fabric to increase thermal resistance still further. Gutlein *et al.* recommend a polyurethane bladder in a textile net for making side-impact bags, while yet another author would like to see needle-punched construction methods used to increase permeability in a controlled manner and describes conditions for achieving optimum gas transfer characteristics.

Medical Textiles

Medical uses of textiles should also not be ignored, as they constitute a major area of advance for the industry.

Industrial Applications

The widespread use of textiles in industrial applications, however, is not generally discussed in any detail. Various authors write articles dealing with the many valuable attributes of technical fabrics, including those intended for space activity, automotive filtration, airbags, composites, marine uses, laminating, seismic protection and resistance to harmful agents.

Applications range widely in type and some consideration of representatives of all the varieties of application should be included. They may be divided into those used directly by industry, those used in leisure or sport activities and those in which heavy duty use is needed in everyday life. Textiles used directly in industry include such diverse items as tarpaulins, filters, acoustic or thermal insulation and protective garments for workers.

Other Uses

Three specific types of more interesting use can be identified. First, transport via heavy lorries is rendered less perilous (and has a lower risk of polluting the road or causing accidents by losing part of the material being shipped) by the use of a new kind of tarpaulin, knitted from stretchable textured polyester, that conforms to the shape of a load, reducing flapping as the vehicle moves.

Second, filters appear to be popular subjects for research, an occurrence caused at least in part by the need for a cleaner environment. Third, leisure or sport applications are rapidly increasing and include items such as tentage, sleeping bags, boat sails, mooring ropes or other marine cordage, aircraft skins, parachute fabrics, climbing ropes, bungee cords, balloon fabrics or guy ropes. Little attention appears to have been paid to most of these applications, although there are some papers worth mentioning in the literature.

Finally, one or two less obvious uses may be mentioned. Newberry reviews new developments in the use of composites in corrosion-resistant applications such as piping, lamp poles, and tanks to contain hot brine or for use in deionised brine service. Risks of harm from mobile phones, reportedly causing brain damage from electromagnetic radiation, are supposedly counteracted by the use of a flexible metal-coated fibre, preferably an acrylic one coated with copper or nickel. Plates used to press boards of plywood, or those for printed circuits, lined with textile cushions designed for high-temperature use, are reported, while Marsh discusses the use of fibre-reinforced plastics for improving under-sea oil extraction; the smooth surface, chemically inert nature and low weight mean that lower losses from corrosion and oil escape will occur, so providing an ecological benefit. Kotliar uses textile and carpet waste to produce a low-cost, wood-like material by incorporating them in a high-modulus phenol/ formaldehyde matrix.

Other heavy-duty uses in everyday life include tyre cords, seat belt webbing, vehicle upholstery, carpeting, wall hangings, conveyor belts, drive belts, tow ropes, isolation suits, protective clothing, space suits, architectural fabrics and geotextiles. In addition, cleaning cloths, surgical clothing or drapes, bandages or other dressings should also be added to the list.

INDIAN TEXTILE

The fate of people of India has been going ups and down and the architecture, music and fine arts and crafts have been growing steadily for centuries. Aliens who came, whether as traders or as invaders, ended with the solitary exception of the English by making India their home and by getting absorbed in its original inhabitants. The intermingling of races did begin new currents and cross currents in the flow of our culture, but it made our culture richer and fuller.

The culture, India invariably shared with its neighbours and others, thereby earning for itself an honoured place-second to none-amongst the nations of the world. Its fame spread to lands far and near, even in those distant ages of which no written historical record exists. Whilst the missionary zeal of our religious emissaries carried the spiritual aspects of our culture to the distant corners of the globe, the spread of our culture on its temporal side was completely because of our traders and exporters who were our cultural ambassadors abroad.

Through the export of its famous textiles the exquisite beauty and fineness of which was almost legendary, ancient and medieval India attained an unrivalled position in the international textile trade, which she maintained till not-so-very-long ago.

According to George Birdwood, India "was perhaps the first of all countries that perfected weaving and the art of its gold brocades and filmy smuslins". Can it be wondered then that the textile products art the Indian carftsmen were eagerly 'and even worriedly looked for and awaited in the markets of the world? How could competitors ever hope to catch up with a country that was thousands of years ahead of them in this art? They could only ban the use of the Indian fabrics within their own country. The Roman Senate was probably the first to pass a law (about 1600 A.D.) prohibiting the wearing of the Indian silk garments by men. A little more than a century later England prohibited by law the use of the Indian calico. This was the beginning of the two-centuries-long unremitting war against India's supremacy in textiles. As is only natural, a crop of legends has collected around the fineness and the delicacy of the fabrics which the Indian weavers produced with the aid of little more than bare human hands.

One such, which is widely believed to be authentic, might be related here. The austere Mughal Emperor Aurangzeb chided his gifted daughter Zebunnisan, for not being properly dressed as her body could be seen through. "But I have no less than seven suits on" protested the lady "one on top of another." "Put on some more then" commanded the Emperor to obscure the transparency. As in Architecture, Music and other Fine Arts; India has its own distinctive contribution in the production of textiles so far as artistic designs, harmonious and beautiful colour schemes and, of course, the fineness of the fabrics are concerned.

Hand spinning and weaving though carried on a small scale and developed gradually as a cottage industry had yet reached a high stage of perfection, even as early as 327 B.C. We catch glimpses of it in the descriptions, sent home by Alexander's soldiers, of the costumes worn by the people of India. Foreign visitors to this country, on their departure, always carried away with them, as priceless treasures, pieces of silks, brocades and muslins. Cargoes, laden with our wondrous fabrics, were invariably welcomed by other countries.

This marvellous technique of the craft which has raised it to the status of a 'Fine Art' has developed through hundreds of generations of craftsmen who formed an exclusive class or guild and handed down from father to son, the knowledge attained and the progress made so that each successive generation of craftsmen added something to it and became more and more efficient, producing better and yet better results.

Another feature of traditional Indian textiles common to all artistic creations of human hands is that, each piece bears the stamp of the individuality of the craftsmen who produced it. Thus, it has a human interest of its own. As the gifted writer Shm. Kamla Dongerkery has said, the conventional textile of India "reveal the background of rich culture, they give artistic shape and form to the ideas and ideals which inspire the lives of the people" and thus provide one of the most reliable hallmarks of the cultural development of the people.

The craft was carried on in the villages of some of the important provinces of India, particularly around the coastal towns and cities. It was not restricted, as a profession, only to men alone, but women too, in their own leisure hours, helped their men in spinning and weaving and thus, added to the family income. Indeed, in some provinces like Assam, tradition needed women to achieve some degree of skill in the arts of weaving and spinning and even today unmarried girts in Assam are not considered quite eligible for matrimony, unless they have acquired some grounding in this useful art. But the craft flourished mostly as a consequence of the patronage extended to it by the royalty and aristocracy, which played a vital role in its development. To suit the tastes of the royal and noble families, the standard production had necessarily to be very high and ever progressive.

Various kinds and varieties of textiles, were produced in different parts of India and thus we hear of Benars and Sural Brocades, Dacca Muslins, Cashmere Shawls, Gujrat Patolas and Bandhanis, each famous for its design, quality and colour. In the following pages, an attempt has been made to present, in bare outlines, the chief features of some of the outstanding traditional textiles of India. Illustrations of several of these fabrics have been added to their descriptions.

It would probably interest our readers to learn that in anum. ber of ancient families are cherished, even today, as valuable heirlooms, some beautiful specimens of our traditional textiles, A good few of these have been presented to some national museums which proclaim to the nation and to the world of Fine Arts "the glory that was Ind".

Dacca Muslins (now the capital of Eastern Pakistan) was, for centuries, synonymous with the finest muslins the world has ever produced by hand or machine. In the words of Dr. Forbes Watson "with all our machinery and wondrous appliances we have been hither to unable to produce a fabric which for its fineness and utility can equal the 'Woven air of Dacca'."

Indeed, the Dacca weavers' Imagic hands produced such exquisitely fine I and delicate fabrics that thee poetic names 'Ab-irawan' (Flowing water), 'Baft-Hawa' (Woven-air), 'Shabnam' (Evening dew) were justifiably given to them. Exhibits in some of our museums 'Prove even today that a yard's width of the muslin could easily pass through a lady's ring.

Then there are legends which have gathered around these fabrics. One of them relates that a five-yard piece of the muslin could be packed in a match box. It is, however, an authenticated fact that a 15 yard piece of 36 inches width 'Mulmul Khas' (Royal muslin) in the reign of the Emperor Jehangir, weighed only 900 grains. Of course 'Mulmul Khas' was the Dacca muslin. The value of Dacca muslins is estimated by the number of warp-threads in a given length of the material as compared with its weight. The greater the length and the number of the threads, with relatively less weight, the higher would be the price.

For example, a yard's width of 'Mulmal Khas', the finest muslin, was known to have 1000 to 1800 threads in the warp, and the weaver took five months to weave 10 yards of the fabric. The weaving of these fabrics could only be done during the rainy season, because for the weaving of such an extremely fine fabric, a humid atmosphere was inevitable.

Saris of Dacca: Upto the starting of the 19th century, the Dacca muslin saris, one of the most artistic and beautiful specimens of handloom textiles, were counted amongst their valuable and cherished possessions by the women of Bengal. Even today, these saris are beautiful in their designs, but alas the art of making fine muslins is no more, and the saris too are not half as fine as they were in the past. The saris ate generally grey, white or black with blue or black designs.

Occasionally, the patterns are woven in with bright coloured cotton, or silver or gold threads. The Dacca muslins with the woven in pattern are called 'lamdani' and the typical designs of flowers or figures used by the Dacca weavers are known as 'Jamdani' pattern. The saris have very bold and large lamdani pattern on the 'Anchal' 'Palloos' (end portion) and the borders. The rest of the sari is usually covered with numerous small bootties. The common motif is the round design bootties, which suggest Chameli (Jasmine) flowers and around these are woven the leaves that recal those of the sweet smelling champak.

When the sprays of flowers are spread all over the sari, it is called a lines, the sari is called 'Terchha'. But when the floral design forms a network which covers the entire field, then the pattern is known as 'Jatar' Sometimes in Jamdani design, the flowers are clustered together-like the settings of jewels, and then the pattern is given the poetic name of 'Panna hazare' (thousand emeralds). If a running floral pattern covers the whole field, the expression Phulwar is used; but if the flowers are large and life-size the Jamdani is caused Toredar.

The borders and Palloo or Anchal (end portion) of saris are generally decorated with distinctive figure designs. The figures chosen represent birds, animals, and human beings. Peacocks or 'mayura' and herons or 'hansa' appear to be popular as bird-figures in the designs of Dacca saris. Also some of the motifs indicate the influence of mythological legends, as well as of the local traditions. The designs are commonly accepted as of Persian origin but many of the designs depict incidents from the Hindu mythology also. The most striking feature of the Jamdani pattern is the skill of the weaver in being able to depict the conception of actual motion in the figures he weaves in.

Many a design presents birds with flapping wings as if they are about to fly away. The outlines of figures are always bold, straight and invariably geometric in clean cut lines. Also the intervening space is so well balanced with lines and flowers that a most delightful effect is produced by the combination. It may well be imagined that the weaving of such masterpieces called for the highest skill and craftsmanship with almost unlimited patience as each design must have involved months of work. An idea of the skill of the weavers may be formed from the following description of the process of weaving.

Famous Chanderi Saris: The muslins woven in Chanderi, a place for due to near Gwalior, have earned a name their fine quality. Chanderi saris are mainly cotton with borders and palloos woven in silk or gold threads. Sometimes mixed threads of silk and cotton are utilised for weaving the fabrics called 'Garbhreshmi'. The palloos of these are very artistically ornamented with gold threads while the ground of the sari is checked, with bootties in the centre of each check-square. The borders are woven with double threads which produce an effect of two colours, one on each side. The saris are woven in nine-yard lengths and are very much valued by the Maha-rastrian ladies.

Barta Kam Khwab, and Abirawan: Kam Khwab is the name given to real gold brocades. 'Brocades' is the expression used by Westerners who at first caued it 'Kin Kab' or 'Cin Cob'. According to their conception, brocades are thick textiles with woven in pattern very prominently thrown up on the surface of the face of the cloth. Resultantly, not only Kalil Kawabs, but other similar textiles also like Buftas, Amrus, and Himrus are known to them by the comprehensive English name 'brocades'.

The poetic name Kam Khwab expresses the dream-like beauty and richness of the fabric. It literally means only a little less (Kam) than a dream (Khwab) or a dream (Khwab) reduced (Kam) to reality. The real Kam Khwabs are woven with pure gold threads and the silk yarn is added to provide a body and as a means for colour illuminations. Silver or gold-plated silver threads are also utilised for keeping down the price but gold is usual. These are heavy fabrics and are commonly used for making palloos, blouses and for men's half sleeve Indian jackets, long-coats (Angarkhas or Achkans), ceremonial robes (chogas) and also later on came to be used for curtains and for upholstering the furniture of Public Rooms and Durbar Halls of Princes.

Handloom Textile of Baluchar Buttedar. Baluchar, a small town near Murshidabad, has become a famous and a highly valued name in the handloom textile history of India. The artisans of the locality produced very artistic figured silk saris known as Baluchar Buffedar. In Baluchar Butfedar, as in Dacca saris, the Pallo os were the most elaborately ornamented portions. The field of the remaining portion of the sari was decorated with small bootties of some floral design or figure design of birds. The special feature of Baluchar Buttedar is that the design used for the ornamentation show a strong influence of Mughal Art, which is famous fore its portraits.

The weaver of Baluchar chose motifs of human figures and the popular 'Toranj' (also called 'Kalka' or 'Guldasta') which is the most popular motif in weaving, embroidery and printing throughout India under its present appellation 'the mango design'. An illustration of Baluchar buttedar displays the richly ornamented palloo and a portion of the field of the sari.

In the design of the palloo, the famous ever popular 'Toranj' are seeri as though these are set in a frame. The border of the frame is again elaborately decorated with pictorial representation of a lady smelling a flower and stated in a sort of a nich. The interspaces are filled with neatly arranged rows of 'Toranj' lined with an outer border of flowering plant.

The border design which is a simple and straight combination of a small toranj and flowering plant is continued for the border of the whole sari. The subjects for portraits were either a lady or a nobleman dressed in Persian dress and holding a flower or riding a horse or smoking. Though the subjects were always of Islamic origin, yet the Baluchar Buttedar were very popular amongst the Hindu ladies. This gesture connotes a complete absence of intolerance existing between the two communities. Later on, probably due to the desire for the patronage of the English bosses of the East India Company the woven-in-pictorial subjects came to include figures dressed in European clothes, and holding instead of the traditional flower a wine glass.

The wonderful art of weaving figured fabrics in Baluchar is lost for ever and a few extinct scattered specimen in some museums are the sad mementoes of the perfection it had achieved.

Pot Thans: These are the brocades in which the major portion of the fabric is in closely woven silk with patterns in gold or silver at regular intervals. This fabric is generally woven in a narrow width of 20" to 30". This is not as heavy or thick as Kam Khwabs but is heavier than the other textiles.

Baftas are also used for blouses and Indian skirts (Lehngas or Damans) and for men's Angarkhas or Ackhans too. Ab-i-rawans. These are silk gauge materials with gold or silver patterns only on some portions. Silk gauze ceries with woven-in gold borders and palloos, are also called Ab-i-rawans. The literal meaning of this poetic name is 'flowing water'. Although brocades are a speciality of Banaras for which it famous yet these are also manufactured at Surat and at Ahmedabad shows an Angarkha made in Ahmedabad brocade. The design in the illustration depict the traditional floral motif. In old brocades motifs representing animals and human figures enclosed within floral borders were in use.

Handloom: Textile of Amms and Himms. Himrus are the famous silk brocades of Hyderabad (Deccan). The State's second largest town-Aurangabad is the chief centre of the art of Himru-weaving. Himru perphaps a derivative of the Sanskrit Him (snow) is a fabric used in winter. The ground is cotton, and silk is used for the brocade on the surface.

The yarn used for weaving Himrus is spun so as to produce, when woven, the effect of a warm soft material like wool. The peculiarity of the Himru is that the silk thread which is used to form a pattern on the surface of the cloth, is carried to the reverse side of the cloth and is collected there in clumsy long loops. This forms a rather loose but soft warm layer. Further, the accumulation of the loose threads on the reverse of the cloth, necessitates a lining to all garments made of Himru cloth. Thus Himru garments make very warm clothing suitable for the cold season. When silk thread is used exculsively for weaving Himru, the fabric is called 'Amru'. Amrus are generally made in Ahmedabad, Surat and Banaras. Himrus are utilised for men's Achkans, Chogas, and for female wear also, e.g., for blouses and Lehngas. For generations, the Nawabs of Surat used a special quality of Himru fabrics for their dresses which was know the 'Nawab's Himru'. These fabrics are also used for upholstery and curtains. Patola is an artistically ortnamented fabric. It is a specimen of wonderful combination of the craft of tie-dying (Bandhana) and weaving. Potola is mostly in use as a wedding sari in Kathiawar and Gujarat. In Java and in Indonesia too the Potola fabric is used for wedding dresses. The fabric is so exquisitely and so highly valued that it is handed down from generation to generation in the family. Women of Gujarat and Kathiawar treasure the possession with pardonable pride.

Palo/a, unlike the other ornamented fabrics, is invariably woven in just the plain weave. The elaborate and intricate patterns which mark the Patola saris are produced by the wonderful art of Bandhana or tie-dyeing. The silkyarn with which Patolas are woven, is first dyed by the Bandhana process before it is put on the loom. The yarns, both warp and weft, are dyed in the lightest of colours. Then they are stretched on the ground, and the dyer proceeds to mark certain portions to connote the lines of the desired design. His wife who helps him in his work, then ties up the marked portions with cotton thread so tightly that the next dye cannot penetrate through to the tied portions.

The yarn is then immersed in dye-baths of the desired colours and shades. The operation of tie-and-dye is repeated many times until all the colours and shades needed for the planned design have been applied to the yarn. The dyer begins with a light colour, passes next to a bright one and applies the dark colour at the very last. Then the weaver starts on his job. The warp threads are arranged in pre-arranged sequence, and the weft is then interlaced one by one very carefully to form the planned design.

The process of producing a Patola is, therefore, a very laborious one and is very complicated too. Meticulous care and a good deal of creative imagination is required for marking the correct portions on yarns for dyeing in different colours of the pattern. A very retentive memory is another essential requisite for registering and recalling accurately the sequence of the coloured threads in the pattern. Thus only a few conventional designs are used for Patola patterns. The following eight are used by the weavers of Pattan as described by Mr. G.V. Patel:

1. Okhar bhat or water creat design. The real name of this design on investigation at Pattern, appears to be akhrot bhat, i.e., the walnut design.
2. Chabri bhat or basket design. Here each enclosure containing an elephant is made up of four quadrants which look as if forming a basket when two of them are taken together.
3. Pan bhat or leaf design. It is said to be the leaf of the sacred pipal tree. (Ficus Religiosa).
4. Chowkhadi bhat or a diaper with a double outline design. Each disper comprises three flowers borne on a stem.

5. Rattan chawak bhat or the cross of diamonds design. It has interspersed diamonds also.
6. Nari-Junjar bhat or dancing girl and an elephant design. It has necessarily a parrot in it.
7. PhultJadi (hat or floral design) is usually enclosed in diapers outlined by a single line. Each diaper contains three flowers.
8. Wagh-Kunjar bhat or tiger elephant design. The animals alternate with each other in the design.

There is one more design which is used for dhoties (the loin cloth worn by men). This design comprises the devnagri alphabet and the forms of the letters follow those of the mantras (hymns) in religious book. Pattan, a place in Kathiawar, is reputed to be the birth-place of Patola. The weavers of Pattan later migrated to Bombay, Ahmedabad and Surat and the making of Patolas began at these places also. Orissa weavers also have adopted the Patola technique for weaving their special fabrics like curtains, bed spreads, odhnis (scarfs worn over the head and draped round the shoulders and waist by women) and saris. The famous Sambalpore saris are woven like Patolas.

Paithani and Pitambar: Paithanis are the beautiful and rich saris made at Pattan or Paithan in the state of Hyderabad (Deccan). These are exquisitely fine fabrics with gauze like texture ornamented with gold patterns woven in the texture of the cloth. The borders and the palloos which are woven separately as gold brocades are sewn on to the sari.

The colour of the sari is usually dark orange, red or yellow, with gold lines arranged in a check or in stripes. The inter-spaces are usually filled In with a figure design depicting a goose with an olive branch in its beak. The borders and the palloos have very striking designs in bright and showy colours such as moss-green, canary yellow, and bright pink. The common motif of the designs is the peacock supporting a big vase with sprays of brilliantly coloured flowers so arranged as to form a Persian cone pattern.

The illustration of a Paithani palloo does not have the favourite peacock in the design, but it depicts the harmonious arrangement of the sprays and the surrounding floral design. The vase with sprays is placed between two pillars joined with the toran (arch). The design is worked in silks of blue, red, and white colours on a field of pure translucent gold. The whole effect is gorgeous and is eminently artistic in its perfect harmony. In olden days Paithanis were generally woven to order for the Royal family and the weaver took months to complete a single piece. The value of a genuine real Paithani ranged between Rs. 2000 to Rs. 3000. In modern times, however, such highly valued and gorgeous paithanis are not woven. Pitambars are bright coloured silks 5 yards in length with gold borders sewn on them. These are worn by men particularly when performing any of the religious rituals. Sari of Jandhanis or Choonaris Bandhans are the colourful saris and odhnis dyed by tie-and-dye process. These are popular amongst the women of Kathiawar. Gujarat, Rajputana and Sindh. Premlata Jayakar in her article on 'The Dyed Fabrics of India in "Marg" refers to Bandhanis in the following words:

This garment is of auspicious type. A symbol of youth and romance, love play and the 'Sohag' (wifehood) of Hindu women. It is a garment of laughter'. Indian women are known for their love for bright colours. Also the tradition and the customs of wearing special colours on different festivals, makes it necessary for them to become familiar with the art of dyeing at home.

Thus apart from the expert professional dyers almost every Indian girl learns by practice a good deal of the art of dyeing and Bandhani work.

Bandhanis differ from Patola as regards the stage at which they are dyed. Like Patolas they are dyed by the tie-and-dye process, which, however, is done after the fabric is woven. The fabric is folded over many times until reduced to a small thick square or a rectangular piece. The piece is then damped and pressed on a block on which a design has been carved. The impressed portions are picked up by the finger nails (the nails are permitted to grow specially for the purpose and are used as a sort of pincers) and are then tied up with cotton thread in a thickness sufficient to resist the dye. It needs training and great skill to pick up all the layers at once and make it crinkle in a particular given manner.

The Bhalldhanari or the woman who does the tieing up work works swiftly and ties up all the impressed portions without cutting the' thread but carries it over from one point to the next. The dyeing process is carried out in the same order as in Patolas, starting with the light colours and finishing with the dark ones. But each time, before a new shade or colour is applied the tieing up process has got to be repeated generally, the designs used are copies of udari worn by a Gujarati girl. Few traditional ones and the practice tieing-up the same design over and over again the Bandharis become expert to such an extent that they are able to dispense with the process of impressing the fabric with the design. The motif of the traditional designs used for Bhandharis presents animals, birds, flowers and dancing dolls. When "elaborate designs are used the Bandhanis are known as 'Gha.' chola'. In some of the expensive 'Gharchola' gold threads are woven in to form checks or squares, and then the designs are' formed in each of the squares by the tie-and-dye process.

The 'Choonaries' are very light fabrics, and the designs for these consist of dots or pin heads irregularly spread all over the field of the cloth. Seldom the dots are grouped together to form a design, and the design is known as 'Ek bundi' (one dot), 'Char bundi' (four dots) and 'Sat bundi' (seven dots).

It is interesting to know that in few parts of Rajputana, e.g., Alwar, professional dyers existed till a couple of decades ago, who could dye even the finest muslin in two different colours, one on each side of the fabric at the modest charge of only annas four a yard. This art too is now extinct but specimens can be found in some museums. Fabrics of Palampores Kalasdar: These are available in rectangular pieces and are popular with Hindus and Muslims alike. The former use them as canopies for the images of their gods and the latter as praying carpets. Those designed for Hindus portray scenes from the Hindu mythology, whilst those intended for Muslims are of Islamic origin. The latter depicts the conventional Mi"hab with panels forming a frame enclosing the 'Persian Tree of Life' complete with birds and the branches and animals resting under its shade.

The craftsmanship and the skill of the dyer is sufficiently borne out by the excellent portrayal of the minutest details with amazing accuracy. The French traveller Bernier who visited India in 1663, during the reign of Emperor Shahjehan, thus describes the fabric which formed the drapings of the Imperial courtyard:

Lined within with those chittes, or cloth painted by a pencil of Masulipatam, purposely wrought and contrived with Palampore such vivid colours and flowers, so naturally drawn, of a hundred many fashions and shapes, that one would have said it was a hanging parterre."

Kalarndar Kabrendar Saris. This is the name given to the hand painted cotton fabrics. They are so called because the artist works out the designs on the material with a fine steel brush not unlike a pen (Kalam). The process is very much the same as utilised for Batik work. The basic principle, namely, resist-dyeing, being common to both. The material is first dyed in pale tpink and then stretched out tight. The artist then traces the outline of the design with his Kalam or fine steel brush dipped in melted wax. The fabric is then dyed deep red and finally washed in hot water to melt away the wax. This produces the design in deep red on the background of pale pink.

Fabrics of Embroidered Type. India is rich in embroidered fabrics. the Indian art of embroidery is an ancient one. Its origin can be traced right back to the Vedic age and it appears to have been well-developed by the time of the great epics. Later on, the influence of incoming races and tribes especially those which migrated from the hills contributed a great deal toward the development of the art along lines which have given it, its present form. "The stitches employed and art conceptions displayed" by these early artists indicate the extent of their knowledge of the art. It may be noted that throughout the mountains and valleys of India, the art is very popular and is assiduously pursued by men and women. The very colourful embroidery produced is not always intended only for the market, but for home use as well.

A few of the better-known embroidered fabrics may now be briefly described. Saris of Phulkar-Phulkari really means flower (Phuf) work (kari). These were the conventional ceremonial shawls worn by the Hindu bride at her wedding when going round the sacred fire with the bridegroom. It is a popular saying in Punjab that when a girl is born in a family the mother, or may be the grand-mother, starts embroidering a Phulkari to be presented to the girl on her wedding day. The women of the Jat community are specialised in Phulkari work. In almost all the districts of the Punjab, wherever this community settled down the Phulkari work qriginated and has flourished.

The peasant women of Rohtak, Hissar, Gurgaon and Karnal are known for embroidering the best Phulkaries. Rohtak is considered to be the home of Phulkari work. The stitch for Phulkari embroidery is the simple darn stitch. The material on which the embroidery is worked is rough Khaddar, but the thread used for the embroidery is silk and the colours used are white, red, yellow or green. The colour of the Khaddar material is invariably red, maroon or brown. The Phulkari motifs generally are of floral patterns, but geometric patterns are also used.

For instance in the Bagh or Shalimar design, the entire material is covered with a geometric pattern. In orthodox Phulkari, however, the same floral pattern is embroidered only at intervals on the cloth, intervening portions of which are left plain. Sometimes, only the borders" are embroidered to a width of 3 or 4 inches and the centre of the material is left plain. This pattern is known as chobes. The choice of the design is wide, but very often they appear to be almost identical becauce of the close and compact stitches 'used. It is remarkable that the illiterate village women choose designs and the colour schemes which have a charm and beauty of their own and which are worked out unerringly by memory.

Saris of Shishedar Phulkaries: These are different from the other phulkaries. The special feature is that the embroidery is embellished with tiny mica or looking-glass (Shisha) discs fixed to the cloth by button-hole stitches, The material embroidered in this way is very often silk, or even satin. Sindh specialises in this kind of phulkaries and is reputed to be its home.

The designs used to indicate the influence of the Punjab as well as Cutch in the kind of stitches used and the colours selected. The phulkari work of Upper Sindh appeares to follow Punjabi Phulkari while that of the Lower Sindh bears a resemblance to the Cutch work. Besides the, button-hole stitch, the darning stitch of Cutch are also used in the Shishedar hulkades.

Roomals of Chamba: The Roomals of Chamba, a state in the Himalayan range, are remarkable pieces of embroidery. Princesses as well as shepherdesses are equally adept in the art which they have adapted as a pleasant or profitable pastime for their leisure hours.

The Roomal is a square piece of botton material of the size of a teapoy cover. The material usually selected is very fine and of delicate texture. The embroidery is worked in double satin stitch which produces the design on both sides of the material and so a Chamba Roomal does not have a right or a wrong side. The motifs for designs are figures, flowers, leaves, forming a sort of frame to set off the central representation of a scene from a mythological story ora legend. It is said of that embroidery that it "imitates the Pahari School of painting". The special characteristic of the embroidery is that it gives a vivid impression of the embroidered figure being in action or in motion, thereby enhancing the artistic value of the Roomals.

Cutch Phulk ari Silk: This is a name provided to the embroidered silk or satin material used for skirts in Cutch. Illustration' of the Cutch Phulkari. The ower portion of the material is embroidered to form a border, and the rest of the material 'is filled with floral or figure designs. The popular figures chosen by the Cutch artist for embroidery are elephants, peacocks and parrots. Majority of the embroidery is done in cnain stitch, but occasionally herring bone stitch is also used for finishing the edge of the border.

Bengali Kanthas: Kanthas are not original fabrics, but are made with lengths of old and practically used up cloth. Several pieces of about the same length are placed one on top of another and the edges of all the pieces are sewn together so as to form a padded or quilted rectangular piece. Then the artist proceeds to depict beautiful patterns or scenes from stories from the epics or well known legends, by means of embroidery done in simple running stitch. The work is very fine and neat and accurately executed completely from memory without the help of any tracing or drawing or any written notes.

The Kanthas are therefore classed as works of art and one has only to look at them to agree. The women of Bengal devote their leisure hours in working kanthas, creating beautiful and artistic fabrics out of worn out clothes. Embroidered Fabrics of Kashmir. Embroidered fabrics of Kashmir are world famous. The Pashminn shawls, the silk saris, the Namdas and the various other silk and woollen artides are praised as works of art. The Kashmir embroidery is called Kaseeda, and the stitches used are the satin stitch, the stem stitch and the loop stitch. The darning stitch is also used. The nerring bone stitch is used for the edges of the finished pieces. The craftsmen, in Kashmir, are generally men and are in variably assisted by boys, often of a tender age, who do the actual embroidery. It is interesting to note the young boys at work. The master craftsman calls but from the design before him, the kind and the number of stitches to be put in.

As the instructions are called out the boys work swiftly and deftly with needles and the stitches are completed almost as soon as the master hastinished calling them out. The scene resembles that of a small, class in a school with the boys taking down the dictation of the teacher. Nature's bounteous gifts appear to have been literally showered on Kashmir and it is considered by several to be the nearest approach to "a paradise on earth".

Kasheeda artists, therefore, under the inspiration of the beauty of the natural surroundings which they have succeeded in reproducing in their embroidery with such amazing skill, that in the, Words of Shm. Kamla Dongerkerry, "competes with the wealth of Nature's charms".' The Kashmir artist has an inexhaustible treasure of motifs in gorgeous colours, ready at hand from which he can draw his designs.

In Kashmir embroidery, therefore, we find bunches of fruits, foliage and birds of brilliant hues and-innumerable shades of colour lend themselves' as charming compounds of various designs. And the Turanj (of the mango) is not ignored either. Indeed, it finds a place in almost all the Kasheedas of Kashmir. The chief features of Kashmir embroidery is its elegance and the harmonious grouping of the brilliant colours in the design which produces a restful and soothing effect.

The 'workman-hip is so neat and the stitches so fine that one wonders how bare human hands and that too of young boys could produce such work. It must, however, be remembered that the training of generations (the art is handed down from father to son in each of the families) has equipped the artisans with almost an intuitive aptitude for the work which has become second nature to them.

On the other and, the art lives, only so long as the line of the craftsmen continue and is likely to become extinct with the line. This, alas, is what started to happen some' time ago due mainly to the rage for replacing human hands by lifeless machinery. Paradoxical as it might seem science is fast stifling and killing art by restricting the scope of arts creation only to such articles as lend themselves to mass production by mechanised processes.

The modern products of Kashmir are no longer as fine and delicate as in the past, and tend to lose their natural charm. The embroidery is best done in silk but the identical patterns are also repeated in wool. Apart from the Pashmina shawls arid the silk saris, the industry produce many articles for use in the home.

Namdas rugs are the embroidered rugs. The rug is a felted thick fabric manufactured by the process of pressing wool and cotton together. They are then embroidered in thick wool in bright colour designs similar to those on Pashmina. Kashmir Silk embroidered fabrics besides bring colourful and rich in design and comparatively inexpensive, are within the means of middle dass people. Thus wider patronage supports the industry and as the art has nourished not by the patronage of the great and the rich alone. Muslins of Chikankari Embroidered.

The white embroidery on white cotton specially on muslins is known as Chikan work. Chikankari is an industry nurtured and developed in the region watered by Ganga and her sister Yamuna (Jamna). Lucknow in special is the cradle and unrivalled centre of the art. Chikan work lack in the attractiveness associated with colour yet has a fascination of its own, unequaled by few and surpassed by none. Whilst the kaseeda of Kashmir may be rich in colour, reflecting in silk the ravishing beauty with which Nature has gifted the valley; the virgin white chikankari is perhaps a translation in shnple cotton of the purity of the waters of the sister rivers born in the home of perpetual snows the Himalayas.

Much credit for the high position of the art of the Kashmir embroidery goes to its attractive colours. The fact that without that colourful aid, the chikankari has attained the same eminence, is eloquent of its great artistic value. Daintiness and delicacy (which are the hall-marks of all Lucknow works of art) added to a firtish and a richness of its own, are the outstanding characteristics of chiktmkari. The work is sometimes so fine that to the naked eye it presents, the appearance of having been woven-in, along with the fabrics.

There are two chief styles of chikan embroidery:

- The knotted or the embossed, of which the murri and the phanda are the well-known varieties. To these may be added a third style, namely, the jali or netting which is akin' to the drawn thread work, but is produced in an completely various 'way.' The drawing out of threads, is, regarded slovenly by the jali embroiderers. 'Instead, the fabric 'is pieced with holes of the requisite size to suit the, pattern and these are then tied up to produce an appearance of net. Jalis, are of different kinds variously named Madras loli, Calcutta Jali, etc. and are very elaborate and intricate. The varieties of designs are said to be thirty in number. Considering the fact that the colours are not utilised to vary the designs, and only the forms and motifs alone produce the diverse patterns, the number is creditable. The design represents familiar objects connected with daily life, often' grains like rice and millet, in a variety of combinations.
- The flat—this group includes the bukhi and the katao styles.

  'The katao produces 'an effect resemble to bukhia. The save fabric is used to produce an opaque effect and for the outline' a simple prdinary stitch is used.

  The bukhia is the most intricate as it is the most important of chikandesigns. It is supposed to be the true chikan. Shri. mati Kamla S. Dongerkerry considers its comparable to the shadow work of the present day" and thus describes the technique of its production:

  "The stitches in this design cover the back of the cloth in the style of the herring-bone stitch, producing an opaque effect on the surface of the fine white fabric and at the same time an out-line of motifs of flowers and leaves' with minute stitches resembling the strokes of the back stitch.

  The murri and the phanda are used mostly in the patterns representing grains. Chikankari work is generally used for sari borders, blouse or kurta collars and tm recent times it was also used for the neck pieces the cuffs, and even the hems 'of men's angrakhas (long, coats). It has lately come into use as table linen, tea-cosy covers, and numerous other washable articles of domestic' use. ' Perhaps due to the influence or' the Western customers who invariably look for colour in Eastern products, the embroiderers have begun to use coloured threads to a little extent which alas deteriorates the pristice elegance one spotless white.

Textile Manufacturing and Use of Robotics

In this, we will find extensive treatment of decision making while the robotics in textile is being contemplated to bring into use. So, first, let us make effort to visualize possible developments that could take place at some time in the future in a most modern yarn manufacturing plant. This plant would be designed and orchestrated to produce high quality sales yarn at a cost which would not only be competitive for American mills, but also for worldwide competition. In this modern yarn manufacturing plant, cotton will be received in uncut bales with classing samples attached and incessant in moisture-proof bags.

These samples will have been removed during the ginning process. Besides the bill-of-lading, each shipment will list not only the bale number, weight, and grade, but also the fiber test results, such as two and one-half percent span length, fibers shorter than one-half inch, fiber strength, fineness and nearly any other feature that may be desired by the buyer.

The weights and fiber tests will be so accurate and the shipper will be so reputable that it will not be essential to grade and sample all bales received. Usually, only a small percentage of bales will be tested in order to verify the quality of the shipment. Each shipment will be unloaded onto a power conveyor that will transport each bale past a receiving station that will automatically register weights, apply the mill identification number; and remove any samples for testing. All samples of the bales and the shipping lists will be sent to a grading laboratory for observation and to verify the quality of the shipment soon after it has been received.

After passing the receiving station, each bale will be moved by the power conveyor to a loading station where it will be picked up by a robot and taken to storage. All bales will be 'stored randomly in racks, each bale individually suspended. The position of all bales regarding bale number, weight and fiber features will be loaded into a data bank for future determination of each bale's use in the blends that will be necessary to produce optimum quality and high manufacturing efficiencies. When a bale is selected for processing, it will be removed from the warehouse by a robot on a first in, first out basis. It will be transported to an opening area where ties and bagging will be removed automatically. The bale will be placed in a staging area for blooming prior to entering it into its proper position in the opening line for blending. Blending and cleaning lines will be resemble to those being used today; however, they will run close to 100 for, of the time and stop only for emergencies. The speed of the feeders will be varied by direct current drives.

There will be no more stopping and starting of the lines as is usually being done today. This will provide for more uniform feeding of raw material at a slower production rate in order to open and clean more efficiently. Before leaving the opening and cleaning area, let us consider the modern handling of salable and reworkable waste. All waste will be gathered as is done today, with each description of waste being handled by individual vacuum tanks. Below each vacuum tank will be a container of bale-size proportions.

This container will be of sufficient strength to with stand considerable internal pressure. As the level of waste in each container reaches the top, the container will be picked up by a robot and transported to a press, into which the container will be positioned. After the container reaches its proper position, the ram of the press will compress the stock and it will be held in a partially compressed state by retainer dogs.

When the ram is withdrawn, the container will be removed by the robot and returned to its original location under the vacuum tank to receive more waste. These steps will be repeated until sufficient waste is in the container to complete a general size bale. Then, it will be strapped, and bagged if essential, during its last trip to the press. This procedure will allow efficient handling of all waste and highly automated bale press operations, with only one press normally being required. Chute feeds that run continuously will feed cards operating in excess of 100 producing high quality webs.

Long-term sliver variation will be so well controlled that the overall coefficient of variation of one-yard lengths will be on the order of two percent or less. This will be accomplished by sensing and controlling tile weight being fed and delivered continuously. With drafting elements located in the calender roll section, manufacturers will be able to spin lower quality yarns directly from card sliver. The cards will be doffed automatically and at random, with specific lengths of sliver.

Robots will be programmed to pick up cans from each card, place them onto a sliver truck (eight cans from eight separate cards on each truck), and transport them to a staging area for drawing for conventional operation. This will give for excellent cross-blending and it will be a simple matter for an operator to creel the cans from the truck into a drawing frame. The drawing frames will run continuously and doff automatically. As previously outlined, one can will be picked up from each delivery for drawing to be placed on trucks which will hold the proper number of cans for the next process.

Lap winder laps will be transported to comber creels on conveyors in sections, in order that a complete comber can be creeled with minimum loss of running time. Each succeeding process will be doffed automatically and, where needed, robots will be used to transport full creel sections throughout the remaining processes in the card room. It will be possible to doff roving frames automatically straight up into spinning creel sections.

The spinning creel sections will be transported to staging areas, with the roving description loaded into data banks. Spinning creels will be electronically summoned immediately prior to the time they are required, transported to the spinning frame sections designated for creel piecing. The length of each spinning frame will be determined not only by ring size and gauge, but also by multiples of the number of spindles on roving frames. For instance, for 96 spindle roving frames, spinning creel sections will comprise roving bobbins for 96 spindles and sides of spinning frames will be in multiples of 96, e.g. 192, 288, 384, etc. Spinning frames will run continuously, with each spindle being doffed and pieced up automatically and randomly. This will be accomplished by an automatic piecer-doffer which will be mounted on and will patrol each side of each spinning frame. Each bobbin will run until it becomes full, at which time it will be stopped, doffed, and the end pieced up robotically.

This will allow for efficiencies approaching very close to 100 per cent (99.93 per cent for a 30-second cycle time when doffing and piecing bobbins that run 12 hours). With the low cycle times that will be developed by the random doffing and piecing of spindles, spinning frame will be designed with more narrow gauge and ring sizes-considerably less than two inches for medium count yarns; this will permit much higher front roll speeds than those normally being run today.

The doffing and piecing robot will deposit bobbins at the end of the frame into doff boxes which, when either full or as required, will be picked up by transport robots and taken to the winding area where the yam will be staged.

The description of each spinning doff will be loaded into a data bank and, when required at winding, the spinning doff once again will be picked up by a robot on a first in, first out basis and transported to a winding loading station. It then will be positioned into a hoist, the doff boy will be elevated, and then will become a hopper, feeding yarn to each winder spindle automatically. The yarn will be cleared while it is being wound, similar to today's practice.

However, we will receive much more useful information from the clearing equipment, which will be able to designate various classes of defects either per unit length or per unit weight. In effect, the electronic clearing system will become an effective laboratory tool capable of monitoring virtually all yarn defects on a 100 per cent basis. Most yam will be wound onto parallel tubes rather than tapered cones due to new developments in creels and storage feeds that do not need tapered packages.

This will allow greater machine efficiency both in winding and in knitting, due to the larger packages that can be produced with fewer winding defects than by today's standard for tapered packages. After winding, all cones will be doffed automatically and transported to an area where they will be packaged by robots, with all cases being strapped, weighed and labelled automatically. All cases will be stored randomly in warehouse areas, with the description of yam in each carton being loaded into a data bank before shipping.

When shipping instructions are received, computers will select cartons of each yarn description to be shipped on a first in, first out basis and, through the use of robots, they will be transported to a loading area. In the loading area, whole shipments will be made up prior to the truck's arrival. Each shipment will be encased in portable containers mounted on rail tracks which lead directly to the door of the shipping dock. When the carrier arrives, ridged connections will be made with the transport track to a track which will be built into the bed of a truck.

After the truck is positioned at the loading dock, the whole shipment will be loaded and ready for departure within five minutes. This will provide for more efficient use of manpower and equipment, both in the manufacturing plant and for the carrier. Computers feeding on-line printers will make up bills-of-Iading and invoices which, with the use of communication modems, will be loaded in computer files in the accounting areas of the manufacturer, the carrier, and the customer.

No invoices or checks will be mailed. Payments will be made electronically by transferring balances from the customer's bank account to the manufacturer's bank account, with updates on all balances, payments, and receipts being made electronically in each office each day. Now what we have described may seem far-fetched; however, in most instances, the technological art at each point of this hypothetical operation is either being practiced today or is at some stage of development.

As you have already recognized, much of the operational and most of the service work done at each process may be handled by robots. The complement of people needed to run this plant will be highly skilled technicians and observer/ operators whose primary responsibilities will be inspection and correction of functions that take place. This is where we come to the most vital and important part of this operation—people. Without adequately motivated, trained and skilled people, no operation can be a success. Very trained employees, skilled in electronics, microprocessors, computers, and enhanced electronic systems will be required to keep this plan operating smoothly and efficiently.

All levels of management, in addition to being thoroughly skilled in manufacturing technology and modern management procedures, must also be needed either of be skilled or at least thoroughly related to electronic and computer systems. Really this represents a challenge to all who expect to be successful in the future.

With a plant so highly automated, we must not overlook the importance of human requirements. It is imperative that we percept these individuals must not only be highly skilled and self-motivated, but they also must be developed through a quality of life that transcends all fragile relations between man and man, and man and machine. How soon may we expect to be able to build and successfully operate a plant similar to that just described? It may be sooner than we think!

Hopefully, we are in a financial position and are operating successfully in an area which permit us install capital equipment as we know that it will be necessary in the reasonable future, either because of quality needs, improving the working atmosphere of employees, cost reduction, or for environmental consideration. Of course, we must not allow ourselves to be lured into a position whereby, with the reduction of direct operators and service employees, we end up with a host of engineers to keep our plant running. Manufacturers of robots and similar equipment must realize that for their product to operate successfully, all parts, which comprise both the electronic and mechanical actions, must be thoroughly and absolutely reliable enough to operate on a continuing basis—24 hours each day.

Excessive down time on equipment of this kind cannot be tolerated. Faults which occur, although they may be caused by high tech difficulties, must be soon repairable by employees who are not so machnically qualified. Personally, I would not want to build operate a plant today as I have just described it, because do not believe it would be totally feasible at this point in time. It perhaps will be practical at some time in the future. Operations in some areas of manufacturing as described are possible today and each of us with open minds definitely should give full consideration to going ahead and "testing the waters", so to speak, not only to improve quality, productivity, and to reduce tost, but also to start developing our people and our minds for future events in our business which may be very resemble to what has been described. The use of robots in textile manufacturing will probably develop in two areas.

One will be for those jobs which are laborious and repetitive the other will be in those operations which make significant improvements in quality and very substantial increases in productivity. Laborious and repetitive jobs which can be hanged by robots in the future are those such as receiving and storing cotton, opening bales prior to blending, transport of stock throughout the plant, and packaging. Operating robots which improve quality and productivity will be similar to those used today in automatic splicing on winders and automatic piecing on spinning frames.

Robots are being used in textiles today and substantial developments are already being made. As we operate in what may usually be described as free enterprise economy, the future use of this new technology will develop as the economics of each area of manufacturing dictate. No doubt about it, all of us can look forward to exciting developments.

Challenges, rather than being reduced, are greater and more enhanced; and certainly, those who succeed will be rewarded by the great future which can be expected for our industry. Aspects of robotic materials handling in textiles. The art of textile manufacturing has advanced from the starting of civilization. It is not surprising that this should have been the case, as life itself was linked to the utility of textiles and it behaved man, at whatever time, to do the best that he could to procure textile materials as he given for himself the necessaries of life and welfare.

By the time of the Bronze Age, looms had been invented that used several rigid materials, such as wood and horn, to position warp yarns in a more or less stationary configuration so that one could more efficiently insert the filling amongst the warps to produce the woven textile. Whorls and spindles had been invented to help in the forming of yarn and its packaging.

The over-all processes of textile manufacturing had become sufficiently institutionalized to where textile manufacture was a major fireside routine engaged in by members of the family, primarily the females.

These textile-making processes involved the transfer of the crude fiber into yarns, initially with the aid of the whorl at the spinning wheel. The yarns then were placed into the hand loom as warps and later onto the bobbins that were loaded in the shuttle, which was thrown backwards and forwards through the yarns of the warp shed to insert the filling into the fabric. The primary power related to this operation involved the human effort of beating up the yarn thus inserted into the fabric which, in turn, was wound up onto the take off device.

As the art of weaving progressed, it soon became apparent that more yarns were needed to supply a loom installation than could be made by a single family. A loom at the hearthside required a large number of daughters constantly engaged in the production of yarn to keep the fabric production from the loom at the level needed by the family requirement. This state of affairs obtained largely up to the inception of the Industrial Revolution. There was, however, a short time when a premature motion towards a factory workplace occurred in ancient Rome.

It was associated with the production of woollen goods mainly for togas, wherein women from various households met collectively in a workplace to produce yarn and woollen fabric for clothing of the citizens of Rome. This movement was interrupted with the fall of the Empire and it required many hundreds of years to elaps before this early movement towards the factory, was to become the generally accepted way of life. The reintroduction, of the workplace at the end of the Middle Ages sparked the renewal of the movement towards the modern industrial society.

As the introduction of a factory workplace was renewed by the Industrial Revolution, the factory quickly became associated with a region with easy access, to power to run axles, which in turn provided some of the energy for yarn twisting and for 100m operation especially filling insertion and the beating up of the filling as it was inserted into the warp sheet. The factory provided the requisite space allocation for yarn production so that a balanced production could be achieved. (In France and other parts of Western Europe, spinning halls were introduced for the particular making of yarn, in which young girls were used as labourers. They kept production rates uniform among a large number of spinning positions by timing their wheels as they sang multipart rounds, such as "Pop Goes the Weasel."

As this extra space was provided, so too were mechanisms for the handling of the yarn and conveying it from a point of manufacture to a point of utilization, that is, the point of insertion into the fabric. While much human labour was still utilized in connection with the early factories of the textile industry, the development of factories occurred in concert with the introduction of non-human handling devices wherever technological ingenuity was sufficient.

It is this feature of the early textile workplace that contained the germ of automation and robotics as we know the processes today. By the time that Jacquard looms had developed to the point where one could produce highly-patterned woven tapestries utilising machinery, the textile manufacturing process had been integrated with what were, in essence, early computers these machines used punched cards to give the complex motions that were imposed upon the 100m heddles. With this programmed control of weaving machinery, figured tapestries and brocades were obtained which, in some cases, closely approximated the finer handwrought works of art that had been produced primarily in the nunneries of Western Europe.

With the expansion of manufacturing processes immediately preceding War II and throughout those period, techniques were perfected for using labour together with advanced machines in the creation of rudimentary "self-thinking" processes. It was a fairly simple matter to devise lathes and grinding machines with human operators for the production of cylindrical shaped objects, such as projectiles and machine screw threads.

A simple extension of this process quickly developed in which the product was armed automatically and then transferred to an inspection station that confimed the precision of the production. On the basis of a "go/no-go" measurement, the product was either accepted within a given channel or, failing the "go" measurement, transferred into a reject channel. It was a simple matter at that point to automatically transfer the product back to the manufacturing site and subject the item to reshaping if that were feasible.

Those items which were produced outside appropriate tolerance levels were, of course, rejected permanently. It was not a difficult matter to extend the go/no-go production inspection scheme to the point where the role of the human operator was progressively minimized to that of overs who ensured that the several automatic setting and conveying steps in the processes were operating properly in their appropriate sequence. From this type of process development, it was a relatively short time until there was developed not only a go/no-go inspection surveillance of the production, but instruments that could be used to measure chemical and/or mechanical disturbances outside of an accepted tolerance. The measurement was used to activate appropriate mechanisms to move the process back within accepted tolerances.

The role of the human was thus removed from the production process, and what many have termed the second industrial revolution was clearly recognized. The second industrial revolution produced devices which replaced not only the muscle of man, as took place at the starting of the first industrial revolution but also replaced man's mental capacity and ability to observe with appropriate nonhuman mechanisms, so that now the production essentially was completely automatic.

A "generalized robot" had been built to replace man's immediate role in the processes of manufacture, In the textile industry, one of the first clearcut examples of this total replacement comprised the recognition of off-specification product quality. The process was equipped with a means for "feeding-back" that recognition to a station which adjusted the controls so as to obtain the needed specification. This early textile process control involving feedback is found in the control of temperature and of size bath concentration, which are used to coat the warp yarns to be used in the 100m.

This kind of process control with its inherent feedback and essentially labour-free character was clearly in place by the end of the 1950s. By this time the, second industrial revolution was gathering momentum throughout America. In the textile sector again, semi-automatic production control, developed in small steps in connection with production of synthetic yarns, provides a second instance of production with feedback control.

This occurred in connection with the control of the viscose dope in the production of viscose filaments. In the art of filament handling, there are a number of processes that involve the manufacturing of filaments after processing to further increase the yam-like character of the filament product. This involves the processes of texturizing which, generally, again involve not only automatic handling, but also a programmable sequence of temperatures that is imposed on the yarn as it is manufactured through the processing equipment in order to achieve the desiled snarled, interlaced, character.

Such machines are elaborate robot factories. They not only handle material and call for various thermal treatments of it, but they can be programmed to produce a wide number of processes, thus giving the textile manufacturer an enormous inventory of yarn-like items made entirely from the synthetic filament technology of the 20th century. Also in the case of filament yarns, a bundle of filaments appropriately interlaced by means of air jets produces a structure that can be subjected to twisting, plying or air texturizing to produce a looped and whorled structure that can be plied, plaited and/or twisted. This yields a continous filament structure that has surface characteristics similar to those of the conventionally spun yam produced from staple fiber. Along with these developments of' filament technology, newer kinds of spinning processes have developed, involving conventional staple products. The open-end spinning process involves the introduction of staple fiber by means of an air stream into a yarn forming chamber that imparts intermingling between the filaments and a twisting of the product to produce a yarn which is resemble to more conventionally spun yarns.

The primary mechanisms in these machines with their robotic characteristics are due to European manufacturers of textile machinery. The starting of these processes involves suction cups or mechanical fingers that can be moved over the surfaces of the fiber bale to progressively "eat" the material off the bale and convey it through air handling passages to the card.

The product of the carding and drawing prodesses is fed into the open-end spinning machinery. All of these examples give rise to a certain perspective concerning the current art in the handling of textile materials. It has attempted to provide not only a retrospective review of various aspects of the production of textiles, but also to indicate that, as the history of textile production has unfolded, the introduction of advanced machinery occurred. Ultimately that machinery was such that it performed in many aspects as if it were "thinking", so as to produce a product without the benefit of essential human labour.

This is a somewhat more complex view of the robot than that usually discussed. Accordingly, one is forced to consider that a highly articulated mechanical device with appropriate sensing elements to allow it to "see" as well as to "think" perhaps is only a simplification of the more modern production techniques which will develop in the future. In the textile industry it is already possible to see places where the robot and the production mechanism may be thoroughly integrated. In the case of certain modern plants the old spinning frame is encircled by a travelling device which has articulated motion and is capable of sensing whether or not a given yarn is broken or a bobbin has been filled.

At these times the device plucks the filled bobbin and places it on a conveying belt to the production output station or ties up the ends which have broken as a result of a failure in the spinning process. Alternatively, mechanical devices for conveying yarn packages to looms can be seen in the automatic guided vehicle small cars moving on a track operated electronically by computers. These are activated when the loom senses that it is about to run out of yarn supply, and an inventory system is tapped automatically. The essentially yarn is placed upon the car, which then is sent in response to the request of the loom station to that loom where the yarn is needed to sustain continued production.

A variation of this process, using much (If the same equipment, is activated when the 100 has produced a fun beam of fabric, at which time the loom is doffed and the finished fabric rolled onto the conveying trolley and transported to the production shipping point. This process can be visualized as integrated into finishing or other subsequent processing plant.

All of this more highly integrated production turns the total plant into a complex robot which consumes raw material at its receiving dock in the form of fiber or filament and manipulates it through mechanical conveyance and sensing stations, the results of which are fed into computers which turn activate extra-conveying and moving devices to manipulate the fiber into yarn and ultimately into fabric. The principal human content of all of this is that of the overseer of the process.

From the economic point of view, this new-labour factor (that of the robot factory) has a fixed cost with a certain maintenance need, and the new human labour requirement involved in the production is now that of a more complex reasoner or decision-maker who monitors the overall activity of the production unit. The opportunity for this decision maker to engage in entrepreneurial activity is comparatively limited, but there still is an essential and major difference in this human labour input, compared with that of the labour that operated factories at the beginning of the Industrial Revolution.

The second industrial revolution has, in fact, separated that which is machine-like from that which is intellectually more stimulating from human labour content; it has permitted the human component to perform in a somewhat more elevated environment and transferred the drudgery of the process to that of a non-human robot slave. The techniques for the production of material things underwent a substantial change at the starting of the first Industrial Revolution. The first revolution can be thought of as the introduction of automatic transfer of power from process unit to process unit conveying work material from machine to machine, or performing automatically special work tasks using a power source which is available in the workplace.

The second revolution culminates in the recognition of alutomatic devices which are programmable to stimulate through feedback loops different aspects of this power transmission to accomplish process modifications upstream of the sensing measurement. In this sense the two industrial revolutions can be thought of as end-points of a scale of production which has been continuously varying throughout history from the first toward the second. The manner in which human labour is utilized has changed substantially. In the beginning it needed the collection of workers from their homes into the factory workplace and the organization of these workers within the scheme of production.

The second revolution leads to a displacement of many of the worker positions in the production process but introduces the need for more highly skilled service employees to keep the process operating at the desired level.

In the earliest phases of robot introduction into textile processes much of the traditional yarn making and weaving process details are maintained in the newer technological scheme, and the automatic devices with their feedback loops are simply embroidered around the existing processes, thereby reducing the number of human labourers that earlier were employed in the chain 6f production. Devices for the automatic retying of yarn ends on spinning frames where breaks have occurred at particular spindles represents one of the simplest examples of the substitution of automatic equipment for human labour.

More recently, robot-like devices with limited articulation have been mounted on carriages which circulate round spinning frames and offer not only automatic retying of broken ends, but also provide doffing and the starting of spinning by the throwing on of yarn onto an empty take up package. In this way the production of yarn can be made entirely automatic from sliver to the wound-up package. It in turn can be conveyed automatically to a next stage in the production chain.

Additional robots are under development that will allow the packaged yarn to be picked up from a supply source and conveyed from the point of yarn manufacture to the point of use within a weaving or knitting process or the like. The robot transfers the yarn, puts it into the next stage of the textile production chain and begins the production process. These motions require substantial amounts of movement through space and a series of sensing devices that control the subsequent motions. Another textile-related aspect of the second industrial revolution is found in the elaborate automation involved in the maintenance of material inventories.

As an example, a picking device can traverse a predetermined set of fiber bales to produce the necessary tufts and convey them aerodynamically to the card for the formation of sliver which is then fed into spinning devices, for instance, open-end spinning processes. The automatic feeding of these devices can be structured in such a way that one can sample statistically the contents of a series of bales so as to achieve the ultimate in blending from a given supply of baled staple fiber.

Similarly in the case of continuous filament yams it is possible to receive at the input of a factory site the large cheeses or straight wound packages, which are transferred to a series of trolleys. When an appropriate encoding is provided at the loading station these convey the yam to an appropriate spot within the inventory area, to be withdrawn on demand. Such production units are now available wherein hundreds of looms can be found on a given weaving shed floor with relatively small amounts of human labour, on the order of ten workers per 300 looms.

The looms are constantly supplied on demand by trolleys working from an inventory containing the necessary yam components for the fabrics being woven within the factory a given period. The human labour content is primarily that of correcting various weaving errors which originate from yam irregularities and providing machine maintenance. In some cases this minimal labour force must sense the approaching requirement of a given machine for input supplies and signal that need to the inventory delivery system.

This type of automatic production of necessity requires either the automatic sensing that a beam has been filled with woven fabric or that the pirn supply for the feeding of the shuttles is running low and needs resupply. All of the processes briefly alluded to above represent textile processes that have developed through the first and second industrial revolutions and thereby contain a minimum of human labour. Further, labour which is inherent to the processes tends to be of a rather highly trained and sophisticated variety.

Individuals must understand the nature of the automatic sensing devices and the feedback loops, and the controls which are passed through such loops, as the processes for self regulation continue to operate throughout the production cycle. Further the complex mechanical and electrical mechanisms which are involved in this produclion system comprise both elementary mechanisms Mid highly sophisticated electronic and electromechanica devices. One can quickly infer from this that the processes for programming these processes are indeed complicated and comprise a complex set of inputs which can and are changed from time to time as the production is tuned towards higher and higher quality levels or towards greater and greater productivity with some acceptable fault level. Sewing industry research due to of the attention being given to Japanese automation activity, an impression exists that United States sewn products and equipment industries are not active in trying to improve productivity through automation.

This is not true since the 1960s there have been a variety of attempts to attack both specific and general joining problems. Joining limp fapric semi-automatically or automatically presents a variety of problems which have been and are a regular challenge to equipment manufacturers and the industry as a whole. It is almost impossible to detail the wide variety of attempts which have been made in this area.

Basically it appears that the Japanese have proceeded to collect patent information worldwide, organize it and make an orderly assault on creating a sewing system. In making this assault, teams of government technicians, university staffs and industry engineers roamed the world collecting basic sewing system information. A high proportion of what they are studying is of U.S. origin. Why then haven't the American engineers -mechanics -innovators succeeded in developing a sewing system? What have they been doing all this time? What do they need to achieve success? Sewing innovation Assembling apparel, shoes, tents, curtains, furniture and auto seat covers are but a few of the many activities in which thread or thread substitutes are used to join seams.

Joining parachutp seams represents a substantial difference in handling techniques than joining a styled outerwear jacket - even though both products may use the same material and kind of joining equipment. As joining activities are investigated, the response to solving problems tends to become product oriented. That is, the problems of assembling a product are fragmented into a series of digestable steps. Ordinarily, production people and industrial engineers solve assembly problems by copying the way they have been earlier solved. Equipment firms provide a quick reference point for establishing sewing production lines.

When production lines are set up under time restraints, the opportunity for creative innovation is limited. The sewn products industries have consistently sought to provide better equipment to satisfy both individual and broad industry needs.

Modifications of equipment developed by the mechanic/innovator to solve immediate problems are often incorporated into later versions of that equipment by the equipment manufacturer. Handsewing, either actual or with machines, which needs highly skilled operators may be perceived as either a problem or an opportunity to seek solutions. Solutions have tended to create specialized equipment, sometimes of limited general usefulness.

The Federal Stitch Catalogue lists a succession of alternative stitch formations in general use. Machine suppliers compound this complexity by adding a number of feed and material handling mechanisms. Needle suppliers provide a wide range of alternate shapes and sizes for use on various products and materials. Accessory suppliers display a wide range of folders and button handlers to improve material handling and joining quality.

Joining is a very complex activity which deals not only with many equipment alternatives but also with problems of seam strength, decorative stitching, and in some cases, the creation of air tight seams. When the complex these of sewn joining are considered in a world of changing fibers, fabrics, wovens, non-wovens, knit and molded goods, it is not surprising that fully automated factories have yet to evolve in the sewn products industries. But, this does not imply that some steps toward a higher level of automation have not occurred. The problem is being studied.

Improvement of Existing Materials

The most frustrating facet of joining parts together develops from the wide variety of materials used in creating sewn products.

A substantial portion of existing equipment activity is devoted to developing or improving existing material handling devices. Almost every study proves that moving and positioning materials to, under, and away from the joining process is the chief labour time-consumer within the sewn industries.

The Japanese Apparel Automation Project selected thirty material pick-up devices for study in their "System J" project. Twenty-one of these devices were patented in the USA, five in West Germany, two in Great Britain and two in Japan. From this study, a new technique consisting of a modified pick-up device and "stiffening of material" was developed. What is important in this study, insofar as the American apparel industry is concerned, is that better than two-thirds of all the inventions considered in this report were developed in the United States.

The Japanese preliminary study represents a relatively small portion of all the handling devices which might be used for bringing together and positioning parts to be sewn, bonded or otherwise joined. Omitted are overhead and belt conveyors, sensors and switches, indexing equipment and many other devices, in full or pertial use within the sewn products industry.

Machine's Position

Matching one part to another in the joining process is being approached by the machine industry in a number of ways. Most of the methods use an operator to feed a clamp or clamping point which in turn actuates a series of positioning and joining actions. These semi-automated activities are can controlled, can and electronic controlled, or electronic controlled.

A the methods use equipment to actuate the joining operations, an operator can generally handle more than one machine. Skill levels are reduced while productivity is increased. In some cases a series of activities are combined in one sewing centre. Pieces of equipment are combined to do more than one sewing operation or to accomplish intricate jobs semi-automatically.

Material Changes

As properties of materials change, the methods of accomplishing product design characteristics also change', while there is nothing new in shaping materials through steaming, pressing, stretching or moulding the characteristics of many materials allow for a broader use of these techniques today. Moulding of products ranges from bra cups through wet suits to automobile seat covers. Several of the moulding or shaping techniques involve adhesive supporting materials to improve joining bonds and to improve retention of shape.

When these techniques are considered as potential displacement of other sewing activities, another range of complexity faces the designer/ production/ engineering activity. Shaping or forming products instead of sewing them has seen broader use over the past 15 years, and is achieving increasing acceptance. Items live slacks, uniforms and dresses have been formed in whole or in part.

Patents have been granted in the United States, Great Britain, West Germany, Japan and Czechoslovakia for making semi-complete and complete garments. Techniques as varied as spot welding by heating a bow on lingerie through using sound to cut buttonholes are beginning to achieve acceptance. Blends of natural and manmade fibers offer the innovative designer opportunities to create new look and better products through the use of moulding and bonding techniques.

Investment Equipment Research

Most sewing products firms are small and tend to produce limited volumes of goods for splintered markets. Small firms with limited management are not likely to make investments in equipment research and development.

The fragmentation of this industry is, in itselt, a barrier to joint equipment manufacturer user efforts. The sewn products industries represent a low investment sector of the economy. Seldom no 'investment is made in equipment; the manufacturer rents it. In this environment of small, diverse users catering to limited markets, high speed and high volume equipment has a doubtful market. Sewn products research and development activity is mainly conducted by the major sewn products firms. This R&D usually takes one or more of four forms: (1) in-house R&D, (2) in cooperation with an equipment firm, (3) employment of a research firm capable of producing a prototype or (4) employing a university.

Two other forms of R&D investigative activity have been sponsored by apparel and equipment firms: (1) the Technical Advisory Committee (TAC) which comprises members of American Apparel Manufacturers Association (AAMA) and (2) the now defunct Apparel Research Foundation (ARF) which consisted of 375 dues-paying members and which received some R&D money from the Government.

Apparel R&D

Apparel R&D tends to be cyclical and specific. Departments are created and funded when business is good. Projects are normally highly product oriented, generally devoted to improving quality, reducing cost or both. A high proportion of these projects are basically short term in nature.

This form of R&D can often lead to a competitive edge in the market place. Generally, the risk of investing in this type of R&D is relatively low since targets tend to be specific, limited in scope and well planned. In the earlier example of separating, picking up and positioning materials, 21 of the patents reviewed by the Japanese were U.S. patents.

Of these, in-house R&D resulted in patents being obtained by five companies Stone Mfg. Co.; Cluett Peabody & Co.; Blue Bell Inc.; Russell Mfg. Co.; and Kellwood Co. One patent was held by a research firm, one by an individual, and the rest (14) by equipment firms. Review of patent records in other areas indicates a similar pattern of activity. It should be kept in mind that several large firms, such as Farah and Haggar, do not usually seek patents yet pursue a reasonably active R&D programme.

Besides several smaller firms whose mechanic/innovator develops a device pass the item to an equipment firm to obtain a production item. This increases the equipment firm's patent volume and reduces the activity attributable to the sewn products industries.

Uses of Research and Development (R&D)

This form of R&D again is generally specifically directed at an opportunity. The user is a major prospect for the equipment firm, but ordinarily the equipment firm also perceives a market beyond the user. Projects developing from user/equipment firm cooperation provide for user testing and user first call on the equipment. Success in these joint ventures has been spotty, with most of it in the area of combining jobs. Perhaps the best known joint venture is that of Union Special and Cluett/ Pfaff in developing semi-automatic assembly of shirts. This venture covered a span of several years and resulted in several innovative machines which are now on the market.

There were some ideas which did not work out at that time, chiefly because the electronics were not available. In the recent past, Kellwood/Union Special, LeeIJet Sew, and Haynes/Union Special provide a sample of firms working together on specific problems ranging from automatic bottom hemming to combining three jobs into one. Both progress and failures resulted. However, most of the projects obtain most of the objectives.

Approach to Research and Development

Several research firms are in the position to investigate an equipment objective, determine its feasibility and produce a prototype. More often than not, this kind of activity has, in one way or another, been jointly financed by several user firms. This approach to R&D was popular during the late 1960s and early 1970s. As an example, Blue Bell, HD. Lee and Levi Strauss jointly worked with Battelle Institute on equipment to full-fell jean seams semi-automatically.

Some positioning and pulling mechanisms were developed which evolved into work aids, but the chief program of producing a smooth full-felled seam semi-automatically was never solved. The 375 members of ARF together with the Government commissioned Arthur D. Little Co. to investigate limp material, handling devices. This project was successful through the prototype level. The patents were part of the public domain. Unluckily, equipment producers who move forward with this system to a production model do not have any protection from competitors. Therefore, the results hoped for were not obtained. Automation Equipment Labs (AEL) approached the sewn products industry seeking backing for a series of automation projects. The firm received quite a bit of backing.

The expense associated with attacking it broad spectrum of joining problems developed into more than the backers felt they could afford. As a result, they dropped out of the semi-joint venture. These instances represent just a few of the joint ventures in which users have attempted to move forward. They do have many things in common: The sewn products industries recognize need for better ways of doing things. They have been willing to devote time, manpower and substantial resources to projects. There is a limit to how much they can afford to spend on R&D and still stay in business.

Support to University Research

Sewn products firms have supported university research on a limited basis. Majority of the industry's support has been directed toward giving equipment and funds supporting courses in apparel activities. Occasionally,' advice is sought on chemical or mechanical problems, generally in conjunction with a specific in-house project. Support of university research tends to derive from individuals within the industry rather than as funded corporate projects.

Each year the TAC Committee reviews the state of the art in some area of the industry. Reports are prepared and given to AAMA members. Basically, the reports are a summation of what the committee has found to be in existence within a particular area. It gives the industry an opportunity to bring itself up-to-date on either equipment or operating systems according to the subject selected for the year. The reports also generally indicate trends and provide management with useful background information.

Role of ARF

The function of ARF was resembled to that of TAC in some respects. During its operation, the foundation collected information on equipment and trends. A quarterly report was issued to its members. ARF also kept track of government activity which might be of interest to the industry. It obtained funds for some research activity from the government.

Among the last proposals it made to the government was one to the Presicfent's Domestic Council on October 30, 1971. This proposal reviewed the state of the apparel industry, the activity of the Japanese apparel industry and proposed a series of counter measures. Among the proposals was a request for funds to develop a preliminary piot plant design. The proposal advised that the plant be designated to make military clothing using experimental techniques. The proposal was declined to accept.

TEXTILE SOILS

Soils come from a number of sources in the environment that textile structures are subjected to during wear and use.

These soils include:

- Solid particulate matter (clays, minerals, soot),
- Oil-borne soils (fats, greases, etc.), and
- Water-borne soils (water-soluble salts, etc.).

Solid particulate matter such as clays, metal oxides, and soot is often mixed with water- and oil-borne soils but can also soil a textile alone through application in the dry state. When applied from the dry state, these solid soils can often be removed by mechanical action such as brushing and shaking. Clays in general are complex inorganic silicates with colour derived from the structure of the silicate.

Oil-borne soils are organic hydrocarbons or related derivatives which are soluble in oils. The aliphatic and aromatic hydrocarbons and fatty acid esters of glycerol are the most predominant oil-borne soils. Less polar hydrocarbons such as mineral oil are more easily removed from textile substrates than are the more polar glycerol esters of fatty acids. Carbon-based matter such as soot is not completely soluble in most hydrocarbon solvents or oils but must be considered an oil-borne soil. These soils can be removed by solvent (dry) cleaning or through emulsification and removal in laundering systems. Water-borne soils are usually water-soluble inorganic and organic salts or natural proteins and starches and can be readily removed by water-based laundering systems. Soils from foods can be oil- and/or water-borne soils, depending on the composition of the particular food.

DETERGENCY AND SURFACTANTS

DETERGENCY AND SOIL REMOVAL

Detergency is a term used to specify the ability of an agent to lift and remove soil from a substrate and to suspend the soil within the cleaning media. Agents which aid directly in soil removal are called detergents or surface active agents (surfactants). Since the term detergent has come to mean complex laundry formulates containing several components, the soil lifting components of such formulations will be referred to as surfactants here to avoid confusion. Surfactants are compounds containing a heophilic (oil-searching) hydrocarbon tail and a hydrophilic (water-seeking) polar head that can effectively aid in wetting of a soiled textile surface, in penetration and removal of the soil from the surface, and in suspension of the soil in the liquid medium.

Surfactants are materials which effectively make the transition between the relatively nonpolar hydrocarbon soil and the polar cleaning medium such as water.

The hydrocarbon tail of the surfactant associates with the surface of the oily soil, whereas the polar head of the surfactant associates with the aqueous medium, thus making a transition from the oily soil to the aqueous media. When oily soil is lifted from a fibre by the surfactant the oily soil-detergent combination is suspended as small particles in the medium through micelle formation.

The low-energy micelle formed must be sufficiently stable to permit its removal in the laundering process. Mineral soils, being partially hydrophilic in nature, undergo a more complex process in soil removal. The soil mixes with the surfactant to form a liquid crystal. Additional surfactant forms a complex micelle which includes myelinic tubes to provide sufficient surface area to remove and stabilise the solubilised soil.

SURFACTANTS

Surfactants are divided into five major classes: soaps, anionic, nonionic, cationic, and amphitricha surfactants. Each contains a hydrocarbon tail and a polar head.

Soap

Soap has been known since antiquity as a surfactant for removal of soil from textiles. Soap is readily made by basic hydrolysis (saponification) of animal fats (fatty esters of glycerol). Soap is the resultant sodium salt of the fatty acids, with the composition depending on the source of fatty acid esters. Soap suffers from one major deficiency as a surfactant: in hard water containing calcium and magnesium cations, the sodium ion in soap is replaced by these multivalent ions to form insoluble salts which cannot act effectively as surfactants.

Anionic Surfactants

Anionic surfactants by definition contain an anion (negative ion) as the hydrophilic head of the detergent and are usually sodium, potassium, or ammonium salts of organic sulfonates or sulfates such as alkylbenzene sulfonates or alkyl sulfates. Anionic surfactants are effective in removal and suspension of oily soil and remain soluble in the presence of calcium and magnesium ions.

For this reason, they are preferred over soap and are the most used surfactant in laundry formulations. The alkylbenzene sulfonates and particularly sodium dodecylbenzene sulfonate are used in such formulations. In the 1950s, foaming problems in water supplies were attributed to these surfactants due to their low degree of biodegradability.

Studies at that time showed that branching of the alkyl group substituted on the benzene ring was responsible for this problem. Reaction conditions for formulation of these surfactants were changed so that the more biodegradable linear alkyl derivative was produced, thereby correcting the problem. Since anionic surfactants tend to foam readily, they are seldom used textile processing.

Nonionic Surfactants

The nonionic surfactants contain a polar head which provides sufficient hydrophilicity to give detergent activity. Polymers of ethylene oxide (called polyethylene glycols or polyethoxyethanols) commonly are used as the polar head attached as the the alkylbenzene or alkyl moiety to form the nonionic surfactant. The hydrophilicity of the ethoxy repeating unit comes from the hydrogen bonding capability of the ether oxygen with water. The nonionic surfactants are used in conjunction with anionic surfactants in some laundry formulations and as wetting agents in many textile dyeing and finishing wet processes.

Cationic Surfactants

Cationic surfactants possess a positive cation and are usually quaternary amine salts. Owing to their high cost, they are less important than anionic and nonionic surfactants in detergent formulations. They are mainly used as fibre wetting agents and as bacteriostats and fabric softeners in selected applications.

Amphoteric Surfactants

Surfactants that have both positively and negatively charged hydrophilic groups within the molecule are referred to as amphoteric surfactants. The detergency of these surfactants varies with pH, and they show bacteriostatic activity at appropriate pH. Amphoteric surfactants are effective leveling agents and aid in controlled diffusion of dyes and finishes onto the fibre.

LAUNDERING AND LAUNDRY FORMULATIONS

LAUNDERING

Laundering is essentially a wet cleaning process s in water solvent in the presence of a detergent formulation. The physical parameters, agitation and temperature affect the ea se and effectiveness of soil removal. Agitation permits the aqueous surfactant solution to flow through the textile structure, conveying the surfactant to the soil, and aids in removal of emulsified soil from the fabric. As the laundering temperature increases, the surface activity of the surfactant solution increases, which in turn increases the ease and rate of soil removal. Although higher temperatures markedly improve soil removal, the maximum temperature that can be used may be tempered by a number of factors, including the stability of the textile and its washfastness. The nature of the impurities in the water has a major effect on soil removal from a textile.

If the water is hard and contains significant amounts of calcium and magnesium salts as carbonates, sulfates, or chlorides, these salt ions will interfere with the soil lifting action of the surfactant unless appropriate water softening agents are added. Dissolved iron salts or the presence of clays, silts, and other colourants can interfere with cleaning, also.

LAUNDRY FORMULATION

Laundry powder formulations or synthetic detergents (often called syndets) are complex mixtures of surfactant and other materials including many of the following (average range of composition in syndets in parentheses):

- Surfactant (10-30 per cent),
- Builders and chelating agents (5 per cent-40 per cent),
- Anti-soi1redeposition agents (0.5 per cent-2 per cent),
- Corrosion inhibitors (5 per cent-10 per cent),
- Foam stabilisers and antifoaming agents (0 per cent-5 per cent),
- Electrolytes and fillers (5 per cent-40 per cent),

- Oxygen bleaches (0 per cent-25 per cent),
- Fluorescent brighteners and colourants (0.1 per cent-1 per cent),
- Bacteriostats (0 per cent-2 per cent),
- Perfumes (0 per cent-1 per cent), and
- Moisture (0-10 per cent).

The composition of the detergent formulation will change with the manufacturer and intended use.

The liquid detergents are aqueous solutions of similar composition to detergent powders with the fo1lowing exceptions:

- The anionic surfactants present will tend to be the more soluble miscible potassium, ammonium, or alcohol amine salts;
- The nature of added foam stabilisers will differ, and
- The amount of builder present will be lower.

A series of specialised product formulations, including enzymes, bleaches and brighteners, water softeners, etc., also are on the market as auxiliary cleaning agents.

Builders

Builders are salts added to a detergent composition to improve the effectiveness of the surfactant present through complexation or precipitation of calcium and magnesium and other multivalent salts. The builders act through complexation (chelation) with these cations to form a stable complex or through reaction with the cations to form an insoluble salt that precipitates from the wash bath. Complexing builders include the sodium polyphosphates (trisodium tripolyphosphate and tetrasodium pyrophosphate), amine carboxylates, citrates, carboxylate polymers, and zeolite ion exchange resins. These builders all complex with calcium and magnesium ions to form water-soluble complexes (chelates) or suspensions that do not interfere with the action of the surfactant.

The polyphosphate builders are the most effective builders but have come under increasing pressure over the last decade due to their role as biological nutrients and contributors to algae growth. The percentage of polyphosphates used in detergent formulations has declined in recent years, but substitutes that are as effective have been difficult to find at a comparable cost. Nitrilotriacetic acid was introduced in the late 1960s as a builder, but adverse factors, including its possible activity as a carcinogen, caused it to be withdrawn from the market. Precipitating builders include sodium bicarbonate and sodium carbonate, sodium sesquicarbonate (a mixture of the two), and the borate salts. These builders provide basicity and react with calcium and magnesium ions to form the insoluble carbonates or borates. These builders are not as effective as chelating builders. After repeated washes they leave deposits of carbonates mixed with soil on the textile being cleaned. They also may decrease the water absorbency of the textile with time.

Anti-Soil-Redeposition Agents

Soil removal is a dynamic process in which suspended soil may be redeposited on the textile as well as removed during the laundering process. Addition of agents with appropriate soilrepelling functional groups inhibits such redeposition. Carboxymethyl cellulose is an inexpensive negatively charged water-miscible polymer that forms a thin deposit or coating on the textile and repels the charged soildetergent micelle. Other polar or charged water-miscible polymers such as polyvinylpyroll idone are particularly useful on synthetics as effective anti-soil-redeposition agents and can be incorporated with carboxymethylcellulose to improve the overall effectiveness of anti-soil-redeposition particularly on synthetic-natural fibre blends.

Corrosion Inhibitors

The basicity and reactivity of ingredients found in laundry formulations lead to attack and corrosion of various metal parts in laundry equipment. To minimize this effect, the sodium silicates are added to the detergent formulation.

Foam Modifiers

Excess foaming during laundering can occur readily due to agitation and can lower the overall effectiveness of soil removal. On the other hand, the consumer views moderate and stable foam formation during laundering as an indication of detergency and soil removal.

Two approaches have been used to provide products which meet both of these concerns. Antifoaming agents such as long-chain aliphatic alcohols, emulsified terpenes (naturally occurring alcohols), and organosilicones are used in conjunction with foaming surfactants to lower and moderate foam formation. The second approach has been the use of detergentlike derivatives that modify and stabilise foaming in conjunction with surfactant. These foam modifiers include monoalkylolamine adducts of fatty acids and their polyethylene oxide derivatives.

Electrolytes and Fillers

Inorganic salts such as sodium sulfate are added to laundry formations to bring them up to uniform cleaning strength and to provide appropriate measurable quantities for addition in laundering by the consumer. These materials may be considered fillers but also are electrolytes in solution that serve to enhance to some degree the migration and action of the surfactant as well as improve the physical characteristics of the product.

Bleaches and Fluorescent Brighteners

Oxygen bleaches such as sodium perborate are often added to enhance the whitening power of the formulation through destruction of colour centers remaining on the fabric. Fluorescent brighteners are added to nearly all synthetic fibres in manufacture to cover yellow colouration through blue fluorescence of these colourless dyes in the light. Fluorescent brighteners added to laundry formulations are mixtures of brighteners which have affinity for all fibre types commonly found in a wash load. Bleaches and brighteners also can be purchased and used separately to enhance whitening of the textile substrate.

Germicides

Biologically active germicides are added to some syndets and are particularly important in low-temperature laundering, where biological agents are not destroyed by heat. The germicides include cationic surfactants and phenol derivatives as well as natural products such as contained in pine oil. Chlorine bleaches al so act as germicides in laundering.

Perfume

Perfumes are added to laundry formulations to mask odours of other ingredients and to convey a pleasant odour which may be suggestive of a natural fragrance or of a clean wash. The perfume has essentially nothing to do with effective soil removal but adds to product aesthetics and aids in consumer acceptance of the product.

Fabric Softener

Fabric softeners are product compositions containing a cationic or nonionic surfactant or alkoxyalkylamide, and they may be applied during the laundering rinse cycle or transferred to the textile during drying from an inert cellulosic or polyurethane substrate.

Liquid softener compositions usually contain:

- Alcohol (0 per cent-2 per cent) (to increase solubility of the softeners),
- Softeners (2-8 per cent),
- Surfactant (0-2 per cent),
- Electrolytes (0 per cent-0.25 per cent),
- Fluorescent brighteners (0 per cent-0.32 per cent),
- Germicides (0 per cent-2 per cent),
- Colourants (0 per cent-0.2 per cent), and
- Perfumes (0 per cent-2 per cent).

The rest is water. Fabric softener components used in dryers are less complex, containing softener with a carrier and perfume in an inert substrate.

Starches

Starches used to give a textile stiffness and body are usually added during the rinse or applied as a spray after washing. Starches include naturally derived starch, starch derivatives, and acrylic polymer emulsions.

Enzymes

In the late 1960s enzyme presoaks and laundry products containing enzymes were introduced. The proteolytic enzymes contained within the products must have a presoak period to be effective. They act as catalysts is speeding the hydrolytic attack of protein and carbohydrate components in soils, breaking them down into more easily removed de composition products. Since oily soils are not readily attacked by the enzymes, their use and effectiveness is limited.

DRYCLEANING

Drycleaning is not carried out under dry conditions at all but rather uses a solvent other than water in the cleaning method. The cleaning is carried out in petroleum hydrocarbon (Stoddard solvent), in a chlorinated solvent (tetrachloroethylene and trichloroethylene), or in a fluorohalocarbon (Freon). Although the solvents will effectively remove saturated or oily soils, the solvent is charged with water plus surfactant to aid in soil emulsification of the more hydrophilic soils.

Tetrachloroethylene is the predominant solvent used in drycleaning in the United States. The drycleaning process involves prespotting by an appropriate method to clean any badly soiled areas on the textile. The textile is immersed in the cleaning fluid, and the fluid is circulated through the textile and then filtered through activated charcoal and diatomaceous earth to remove impurities. After a certain period of continued use, the drycleaning solvent is redistilled to remove residual oils and so forth and then recharged with water and detergent.

Drycleaning solvents cause less fibre swelling and deformation and have less tendency to remove dye from the textile. Less agitation is involved than in laundering, and drycleaning is therefore preferred for textiles in which water-induced dimensional change will occur. Care must be taken in drycleaning some textiles, since damage may occur due to attack of the fibre, of finishes on the fibre, or of one or more components in the fibre structure.

MAINTENANCE

A plan to do maintenance work in the future is usually of two types:

- Scheduled maintenance
- Preventive maintenance.

SCHEDULED MAINTENANCE

Scheduled maintenance is that type of work that requires longer durations to complete, planning of manpower and tools and materials required co-ordination with other trades and possibly outside contractors. The timely replacement or maintenance on a major piece of equipment could involve shut downs of other departments or blocks of guest rooms. Projects such as building of walls or complete painting of areas could also come under scheduled maintenance. Indeed, any project requires scheduling and planning.

PREVENTIVE MAINTENANCE

Preventive Maintenance, as it's name implies is the intent to perform timed inspections, minor adjustments, lubrication based on manufacture's recommendations with the ultimate goal of preventing unscheduled breakdowns and prolonging the life and efficiency of the equipment. During the course of the inspection if it is determined that major work may be required, then work orders are generated to schedule the maintenance.

ROOM MAINTENANCE

Room Maintenance, both guest and meeting rooms again follows the above with inspections and generating work orders to schedule and correct deficiencies. The frequency of inspections should be determined to happen sometime before the area slow periods. If guest room's occupancy is peak in summer then schedule the inspection just prior to the downturn as it will give time to order necessary materials and schedule the labour to accomplish the tasks. The importance of inspections cannot be over emphasized because you have yet to see room attendants or banquet staff adequately report deficiencies.

BREAKDOWN MAINTENANCE

Breakdown maintenance can be both negative and positive. Negative if it has an impact on guest comfort, safety, or is detrimental to the smooth flow of production that keeps other departments operational. Breakdowns can be very expensive if it happens after hours and outside contractors are required, or if say the main chiller shuts down and all your guests walk out. Positive as you would not want to spend 100 a year on preventive maintenance on a blender worth 50. Also, in maintenance repairs don't waste 20 worth of time to repair something only worth 1000.

TYPES OF MAINTENANCE

EXPLANATION

Maintenance is something that never ends in a hotel and just when one thinks it is under control either something expensive breaks or it is time to remodel or upgrade!

PREVENTIVE AND BREAKDOWN MAINTENANCE

Maintenance management distinguishes between preventive maintenance and breakdown maintenance. While preventive maintenance can be planned, the breakdown maintenance is invariably unplanned. As regards plant preventive maintenance, the experience with oil refineries, industrial blast furnace, dock facilities and thermal projects have all shown that a thorough planning and a scheduling of preventive maintenance can be immensely time-saving and cost-reducing. The same is applicable in case of tourism transport operations and hotels maintenance, vehicle breakdown or bad maintenance in hotels can adversely affect the business.

Transaction-processing systems have looked at the records of components, spare-parts, jugs, fixtures and tools as inputs, and consumption statements and maintenance accounts as outputs. These inputs can still be valid for building up a corporate MIS with such additional data as equipment conditions, history of failures, direct cost of maintenance, inventory values and materials movement, man-hours spent, over-time paid, other resource usage, maintenance of employee's performance, reliability and maintainability of equipment, and maintenance contribution to the finances.

From such a comprehensive database, one can generate MIS reports for transport control, such as vehicle register (to provide vehicle specifications and project details), vehicle history (to provide maintenance particulars), major failure report, forecast on maintenance (providing probable failure time and period), office non-availability, reliability and maintainability, and maintenance schedules (providing maintenance activities to be done during a specific period).

One can also generate MIS reports on work control, such as, craft performance report (providing planned and actual performance of each craft), craftsman's performance report, maintenance planning efficiency report (providing an efficiency ratio on planned activities), overtime report, resource levelling report (providing analysis of resource required job-wise), delay cause report, etc.

One can finally generate materials control reports and cost control reports providing management details of costs and material transactions involved in maintenance management.

CONTRACT MAINTENANCE

Contract maintenance is mandated in some instances such as for elevator service, kitchen hood exhaust cleaning and fire systems. The reasoning behind this is to ensure that the work is performed by qualified technicians, and may also require licenses and special knowledge. Local regulations and insurance companies usually require these contracts.

It also serves the purpose of making sure the work gets done irregardless of budget restraints. Maintenance contracts or contracting out is almost always necessary to complement an engineering department that is undersized.

MAINTENANCE DECLINING EFFICIENCY

EFFICIENCY – STAFF PRODUCTIVITY

Basically, for the hotels engineering and maintenance staff, the day-to-day operation and maintenance of a large hotel is time-consuming. Therefore, one of the striking advantages of Integrated Building Solutions (TBS) is the complete harmonization of the hotel's HVAC (heating, ventilation and air conditioning), fire, intrusion and CCTV (closed-circuit television) systems.

The integration of all of these disciplines means improved functionality for the hotel, and centralized display of all building information at the touch of a button. Furthermore, TBS also includes voice evacuation capabilities and controls for the hotel's phone system, guest services and background music. In total, the integrated TBS reduced significantly the workload of all involved maintenance staff.

At time when every penny counts, PMWorks, the preventive-maintenance solution from MTech, based here, is streamlining property efficiencies and protecting long-term capital investments for hotels. MTech is a leading provider of guest-response and preventive-maintenance solutions for the worldwide hotel industry.

The hotel management company is one such client. "We welcome Hospitality to our PMWorks family," said Luis Segredo, MTech president. "Management companies looking to grow, particularly with a diverse portfolio, need tools to help them manage the growth and justify it.

With PMWorks, Winston Hospitality can show owners how they are protecting their properties and saving them costly repair dollars. As a subscription service that has been proven across many brands, PMWorks can meet Winston Hospitality's needs at the property level and provide the corporate office with operational and strategic reports.

"Co. continues to add management companies to its growing list of PMWorks users, even in this economic climate, because staffing cuts are requiring automation to help keep service and product levels high," he added. Michael president of Winston Hospitality, says that the proactive nature of PMWorks helps keep their hotels looking in great shape.

"We use PMWorks for everything from plumbing and paint touch-up to case goods and the HVAC system," Heeden said. "HVAC is really important because if it were to malfunction, it affects air quality and the noise is the number one complaint from guests."

Brand quality-assurance marks most often fall in the "outstanding" range for Hospitality properties. Heeden anticipates property mechanical systems will see fewer repairs and have a longer shelf life if each hotel follows the planned preventive-maintenance programme and PMWorks schedule.

WORKING

The Accounts Manager and liaison officer, explained that PM Works enables management companies to better manage the scheduling processes.

"The system provides a work load balanced schedule that lets you focus on the work that needs to be done," Edson said. "Time is no longer wasted preparing a complex schedule that must include critical reminders for Life, Health and Safety. They are all there, in the system. It's a cost-effective solution for management companies that demand corporate-driven quality and consistency across a multi-branded portfolio."

STEPS TO IMPLEMENTING WORKS AT HOTEL

- Enroll the property by faxing or emailing a signed agreement.
- Complete the Web data collection wizard. This is a simple and fast "prompted" collection of data. The hotelier is asked what the equipment is called (Ice machine One) and where it's located. A list of guestroom numbers completes the set up. On average a 100-room property will take about two or three hours to set the entire programme up.
- Data is sent to a regional person, or back to the GM for a quick review to make sure the data is complete.
- The following Monday the hotel will receive its first work order packet VIA email.

When the schedule is in place, hotels receive weekly service orders via email. Each service includes step-by-step instructions on how to do the job right the first time. As technicians complete tasks, they close orders from any touch-tone phone or Web browser. Every week new orders are sent to the property. Included are tasks not completed during previous weeks so nothing falls through the cracks? The management company receives an email showing what properties are keeping up and which are falling behind.

"We receive status reports [green, yellow or red] to see which tasks are completed and which are not," Heeden said. "PMWorks customizes the process to fit the needs of each hotel. It generates the work orders once a week and let's us know where we are in the process. We use PMWorks in all of our hotels, have it ready to go at our properties under construction, and plan to use it in all of our new hotels. "It takes a lot of the guess work out of the PM programme and keeps us on top of critical Life, Health and Safety PM's," he added.

The Hospitality offers services including management, development, rehabilitation, repositioning, and hotel ownership. Its senior management team brings years of experience and a wealth of knowledge to each of these areas and is committed to ensuring the satisfaction of guests and its partners. Its current portfolio includes four hotels, and at press time it was due to open one more.

REPLACEMENT POLICY OF HOTEL EQUIPMENT

PURPOSE

To establish a (PC) equipment replacement and upgrade policy for the Youngstown Metropolitan Housing Authority {CO.}rder to utilize the benefits of next-generation office environments, simplify technical support issues, and increase CO.ility to deploy new solutions to business problems.

STATMENT

Guidelines and procedures are required to maintain a replacement cycle of personal Hotel equipment within the useful and expected lifetime of the equipment, while preventing a proliferation of aging, obsolete, out-of-warranty, unsupported, and incompatible systems.

CYCLE OF REPLACEMENT

It is Co. policy to cycle replaced computers through the major classifications of technology users PCs that have been replaced and are no longer appropriate for one classification of technology users should be used to upgrade and replace a PC with older technology within the agency. In this manner, the PC with the oldest technology in the agency is phased out and the agency can continue to benefit from their investment in PCs.

REPLACEMENT RATIONALE

Hotel equipment, by their nature are relatively inexpensive Part and devices that have a limited life compared to other office equipment. *The rule of thumb for obsolescence of* Hotel equipment *is 3-4 years.* However, changing business practices, new technology, and new software applications can impose increased demands on computing power that can force a more frequent replacement cycle for employees affected by the changing business practices or those using the new technology or software.

REPLACEMENT VERSUS UPGRADE

It is Co. policy to replace PCs with the latest technology rather than to apply major upgrades to existing PCs.

Current research indicates that:

- Upgrading is expensive because, more often than not, multiple system components must be upgraded in order to achieve the intended performance improvement.
- The purchase price of the various hardware components needed to accomplish an upgrade will exceed the purchase price of a new el equipment The cost of labour to install an upgrade to existing hardware will far exceed the cost of labour to install a new PC./ equipment

- Upgraded PCs have uncertain maintenance profiles and support costs.
- There is no increase in residual value of upgraded PCs.
- Upgrading PCs tend to increase the overall complexity in the installed hardware base.

CLASSIFICATIONS OF TECHNOLOGY USERS

There are three major classifications of technology users defined as:

LEADING EDGE TECHNOLOGY USERS

Those individuals within the agency who need to be at the forefront of technology.

Those individuals carry out but not limited to the following functions:

- Heavy usage of the CCS application (processing, programming, upgrading and maintenance)
- Heavy usage of database applications (Access)
- Heavy usage of the Internet (downloading virus fixes, programmes, HUD reports, uploading data to various HUD organizations, viewing HUD reports in PIC, research and analysis)

POWER USERS

Those individuals within the agency whose daily business operations require heavy utilization of information technology. They have a standardized suite of applications on their desktops; customized applications beyond the standardized suite; may use servers for database and decision-support; and often have access to department-specific applications, e-mail, Intranets and the Internet.

Functions performed by these users are(but not limited to) the following:

- Moderate to heavy usage of the Microsoft Office suite (Excel, Word, Power Point, Publisher, Outlook, Access)
- Moderate to heavy usage of Adobe Acrobat and other applications used to complete grants, reports and forms for HUD
- Moderate to heavy usage of the Internet (uploading data to HUD, e-mailing, viewing HUD reports in PIC and other HUD areas, purchasing and reservations)
- Occasional to heavy usage of the CCS application

MAINSTREAM TECHNOLOGY USERS

Those individuals within the agency whose daily business operations require moderate utilization of technology for everyday office functions.

They have a standardized suite of applications, access to file and print servers, and access to some host-based application using terminal emulation programmes. Functions performed by these users are(but not limited to) the following:

CONSERVATIVE TECHNOLOGY USERS

Those individuals within the agency whose daily business operations require the utilization of technology for everyday office functions only. They have a standardized suite of applications, access to file and print servers, and access to some host-based application using terminal emulation programmes. *Functions performed by these users (but are not limited to) are the following*:

REPLACEMENT SCHEDULE

The following is a general guideline for replacing PCs. Departments should consult with "IS" for assistance in determining their specific replacement needs. Replacement schedules vary according to the four major classifications of technology users:

Leading Edge Technology Users

- Should maintain no more than two generations of technology.
- Should consider replacing PCs every second or third year, depending on applications in use.

Power Users

- Should maintain no more than two generations of technology
- Should consider replacing PCs every three to four years

Mainstream Technology Users

- Should maintain no more than two generations of technology.
- Should consider replacing PCs every four to five years.

Conservative Technology Users

- Should maintain no more than two generations of technology.
- Should consider replacing PCs every five years.
  - "IS" is not responsible for the installation and reconfiguration of "unsupported" software packages.
  - "IS" will NOT transfer the entire contents of a user's hard drive. Users who have not properly prepared their machine before their scheduled "upgrade installation date/time" will automatically be placed on the waiting list for future installations.
  - "IS" will NOT be responsible for the installation and reconfiguration of "unsupported" peripheral devices or devices that were installed by users. Equipment originally installed by users will be the user's responsibility to re-install on their new system.

REGULAR AND PROPER MAINTENANCE OF THE HOTEL BUILDING

To ensure regular and proper maintenance of the hotel building and its various machines installed in its premises and to ensure proper hotel functioning, an Engineering or Maintenance departments is there in every hotel.

Qualified engineers along with necessary staff are appointed for this work. These staffs are in fields of electrical, mechanical and civil. The engineers look after the boilers, hot water supply, air-conditioning, plumbing etc. Electricians ensure that the electrical appliances and gadgets are perfectly in order.

Engineering is responsible for the maintenance and repair of the physical plant and all of its features, including:

- Day-to-day system operation and maintenance routines
- Preventive maintenance
- Minor and major repairs and improvements
- Support for other operating departments and guest activities

As the facility is our "product" and the means through which we deliver our valuable services, high performance in this function is critical to the overall, long-term success of the business.

The department plays a critical role in maintaining a safe environment for building occupants and compliance with various federal, state and local environmental, health, safety, operating and building codes, standards and regulations. This department head position carries out managerial processes including budgeting, forecasting, labour scheduling, personnel administration, work order tracking and various record-keeping duties.

The successful candidate is self-starting, proactive, a team player, able to grasp "the big picture" as well as being involved in the details, and is an effective trouble-shooter and problem-solver. The individual must be technically familiar with the many aspects and disciplines involved in constructing and operating a large, complex facility, while effectively performing as a member of the management team. Experience with developing and managing complex improvement projects is important.

Successful Engineering department managers have often been promoted to positions of greater responsibility after initially hiring into the department at the craft level. By acquiring additional skills and demonstrating appropriate qualities, the craftsperson may be given the responsibility of a lead craftsperson or supervisor as an opportunity arises, then eventually managerial duties if further developmental progress is made.

Other managers transfer their proven skills and experience from facility-related positions in other industries or from the construction industry. Those who make the commitment are rewarded by a fast-paced, interesting and challenging technical career within an attractive business environment.

2

Important of Clothing and Textiles

Clothing and textiles have been enormously important throughout human history—so have their materials, production tools and techniques, cultural influences, and social significance. Textiles, defined as felt or spun fibers made into yarn and subsequently netted, looped, knit or woven to make fabrics, appeared in the Middle East during the late stone age. From ancient times to the present day, methods of textile production have continually evolved, and the choices of textiles available have influenced how people carried their possessions, clothed themselves, and decorated their surroundings.

Sources available for the study of the history of clothing and textiles include material remains discovered via archaeology; representation of textiles and their manufacture in art; and documents concerning the manufacture, acquisition, use, and trade of fabrics, tools, and finished garments. Scholarship of textile history, especially its earlier stages, is part of material culture studies.

PREHISTORIC DEVELOPMENT

Interest in prehistoric developments of textile and clothing manufacture has resulted in a number of scholarly studies since the late twentieth century, including *Prehistoric Textiles: The Development of Cloth in the Neolithic and Bronze Ages with Special Reference to the Aegean*, as well as *Women's Work: The First 20,000 Years: Women, Cloth, and Society in Early Times*. These sources have helped to provide a coherent history of these prehistoric developments. Evidence suggests that human beings may have begun wearing clothing as far back as 100,000 to 500,000 years ago.

Genetic analysis suggests that the human body louse, which lives in clothing, may have diverged from the head louse some 107,000 years ago, evidence that humans began wearing clothing at around this time.

Possible sewing needles have been dated to around 40,000 years ago. The earliest definite examples of needles originate from the Solutrean culture, which existed in France from 19,000 BC to 15,000 BC. The earliest dyed flax fibers have been found in a cave the Republic of Georgia and date back to 36,000 BP. The earliest evidence of weaving comes from impressions of textiles and basketry and nets on little pieces of hard clay, dating from 27,000 years ago and found in Dolni Vestonice in the Czech Republic.

At a slightly later date (25,000 years) the Venus figurines were depicted with clothing. Those from western Europe were adorned with basket hats or caps, belts worn at the waist, and a strap of cloth that wrapped around the body right above the breast. Eastern European figurines wore belts, hung low on the hips and sometimes string skirts. Archaeologists have discovered artifacts from the same period that appear to have been used in the textile arts: net gauges, spindle needles and weaving sticks.

ANCIENT TEXTILES AND CLOTHING

The first actual textile, as opposed to skins sewn together, was probably felt. Surviving examples of Nålebinding, another early textile method, date from 6500 BC. Our knowledge of ancient textiles and clothing has expanded in the recent past thanks to modern technological developments. Our knowledge of cultures varies greatly with the climatic conditions to which archeological deposits are exposed; the Middle East and the arid fringes of China have provided many very early samples in good condition, but the early development of textiles in the Indian subcontinent, sub-Saharan Africa and other moist parts of the world remains unclear. In northern Eurasia peat bogs can also preserve textiles very well. Early woven clothing was often made of full loom widths draped, tied, or pinned in place.

Ancient Near East

The earliest known woven textiles of the Near East may be fabrics used to wrap the dead excavated at a Neolithic site at Çatalhöyük in Anatolia, carbonized in a fire and radiocarbon dated to c. 6000 BC. Flax cultivation is evidenced from c. 8000 BC in the Near East, but the breeding of sheep with a wooly fleece rather than hair occurs much later, c. 3000 BC

. But Çayönü in Turkey has also been claimed as the site of the oldest known cloth, a piece of woven linen wrapped around an antler and reported to be from around 7000 BC.

Ancient India

"Cotton has been spun, woven, and dyed since prehistoric times. It clothed the people of ancient India, Egypt, and China. Hundreds of years before the Christian era cotton textiles were woven in India with matchless skill, and their use spread to the Mediterranean countries. In the 1st cent. Arab traders brought fine Muslin and Calico to Italy and Spain. The Moors introduced the cultivation of cotton into Spain in the 9th cent. Fustians and dimities were woven there and in the 14th cent. in Venice and Milan, at first with a linen warp. Little cotton cloth was imported to England before the 15th cent., although small amounts were obtained chiefly for candlewicks. By the 17th cent. the East India Company was bringing rare fabrics from India. Native Americans skillfully spun and wove cotton into fine garments and dyed tapestries. Cotton fabrics found in Peruvian tombs are said to belong to a pre-Inca culture. In color and texture the ancient Peruvian and Mexican textiles resemble those found in Egyptian tombs."

Ancient Egypt

Evidence exists for production of linen cloth in Ancient Egypt in the Neolithic period, c. 5500 BC. Cultivation of domesticated wild flax, probably an import from the Levant, is documented as early as c. 6000 BC Other bast fibers including rush, reed, palm, and papyrus were used alone or with linen to make rope and other textiles. Evidence for wool production in Egypt is scanty at this period..

Spinning techniques included the drop spindle, hand-to-hand spinning, and rolling on the thigh; yarn was also spliced.. A horizontal ground loom was used prior to the New Kingdom, when a vertical two-beam loom was introduced, probably from Asia. Linen bandages were used in the burial custom of mummification, and art depicts Egyptian men wearing linen kilts and women in narrow dresses with various forms of shirts and jackets, often of sheer pleated fabric.

Ancient China

The earliest evidence of silk production in China was found at the sites of Yangshao culture in Xia, Shanxi, where a cocoon of bombyx mori, the domesticated silkworm, cut in half by a sharp knife is dated to between 5000 and 3000 BC. Fragments of primitive looms are also seen from the sites of Hemudu culture in Yuyao, Zhejiang, dated to about 4000 BC. Scraps of silk were found in a Liangzhu culture site at Qianshanyang in Huzhou, Zhejiang, dating back to 2700 BC. Other fragments have been recovered from royal tombs in the Shang Dynasty (ca. 1600 BC – c. 1046 BC).

Under the Shang Dynasty, Han Chinese clothing or Hanfu consisted of a *yi*, a narrow-cuffed, knee-length tunic tied with a sash, and a narrow, ankle-length skirt, called *shang*, worn with a *bixi*, a length of fabric that reached the knees. Clothing of the elite was made of silk in vivid primary colours.

Ancient Japan

The earliest evidence of weaving in Japan is associated with the Yayoi period from about 300 BC to 250.

The Textile Trade in the Ancient World

The exchange of luxury textiles was predominant on the Silk Road, a series of ancient trade and cultural transmission routes that were central to cultural interaction through regions of the Asian continent connecting East and West by linking traders, merchants, pilgrims, monks, soldiers, nomads and urban dwellers from China to the Mediterranean Sea during various periods of time. The trade route was initiated around 114 BC by the Han Dynasty, although earlier trade across the continents had already existed. Geographically, the Silk Road or Silk Route is an interconnected series of ancient trade routes between Chang'an (today's Xi'an) in China, with Asia Minor and the Mediterranean extending over 8,000 km (5,000 miles) on land and sea. Trade on the Silk Road was a significant factor in the development of the great civilizations of China, Egypt, Mesopotamia, Persia, the Indian subcontinent, and Rome, and helped to lay the foundations for the modern world.

Classical Antiquity

Dress in classical antiquity favored wide, unsewn lengths of fabric, pinned and draped to the body in various ways.

Ancient Greek clothing consisted of lengths of wool or linen, generally rectangular and secured at the shoulders with ornamented pins called fibulae and belted with a sash. Typical garments were the peplos, a loose robe worn by women; the chlamys, a cloak worn by men; and the chiton, a tunic worn by both men and women. Men's chitons hung to the knees, whereas women's chitons fell to their ankles. A long cloak called a himation was worn over the peplos or chlamys.

The toga of ancient Rome was also an unsewn length of wool cloth, worn by male citizens draped around the body in various fashions, over a simple tunic. Early tunics were two simple rectangles joined at the shoulders and sides; later tunics had sewn sleeves. Women wore the draped stola or an ankle-length tunic, with a shawl-like palla as an outer garment. Wool was the preferred fabic, although linen, hemp, and small amounts of expensive imported silk and cotton were also worn.

MEDIEVAL CLOTHING AND TEXTILES

The history of Medieval European clothing and textiles has inspired a good deal of scholarly interest in the twenty-first century. Elisabeth Crowfoot, Frances Pritchard, and Kay Staniland authored *Textiles and Clothing: Medieval Finds from Excavations in London, c.1150-c.1450*. The topic is also the subject of an annual series *Medieval Clothing and Textiles* edited by Robin Netherton and Gale R. Owen-Crocker of Anglo-Saxon Culture at the University of Manchester.

Byzantium

The Byzantines made and exported very richly patterned cloth, woven and embroidered for the upper classes, and resist-dyed and printed for the lower. By Justinian's time the Roman toga had been replaced by the tunica, or long chiton, for both sexes, over which the upper classes wore various other garments, like a dalmatica (dalmatic), a heavier and shorter type of tunica; short and long cloaks were fastened on the right shoulder. Leggings and hose were often worn, but are not prominent in depictions of the wealthy; they were associated with barbarians, whether European or Persian.

Early Medieval Europe

European dress changed gradually in the years 400 to 1100. People in many countries dressed differently depending on whether they identified with the old Romanised population, or the new invading populations such as Franks, Anglo-Saxons, and Visigoths. Men of the invading peoples generally wore short tunics, with belts, and visible trousers, hose or leggings. The Romanised populations, and the Church, remained faithful to the longer tunics of Roman formal costume.

The elite imported silk cloth from the Byzantine, and later Moslem, worlds, and also probably cotton. They also could afford bleached linen and dyed and simply patterned wool woven in Europe itself. But embroidered decoration was probably very widespread, though not usually detectable in art. Lower classes wore local or homespun wool, often undyed, trimmed with bands of decoration, variously embroidery, tablet-woven bands, or colorful borders woven into the fabric in the loom..

High Middle Ages and the Rise of Fashion

Clothing in 12th and 13th century Europe remained very simple for both men and women, and quite uniform across the subcontinent. The traditional combination of short tunic with hose for working-class men and long tunic with overgown for women and upper class men remained the norm. Most clothing, especially outside the wealthier classes, remained little changed from three or four centuries earlier.

The 13th century saw great progress in the dyeing and working of wool, which was by far the most important material for outer wear. Linen was increasingly used for clothing that was directly in contact with the skin. Unlike wool, linen could be laundered and bleached in the sun. Cotton, imported raw from Egypt and elsewhere, was used for padding and quilting, and cloths such as buckram and fustian.

Crusaders returning from the Levant brought knowledge of its fine textiles, including light silks, to Western Europe. In Northern Europe, silk was an imported and very expensive luxury. The well-off could afford woven brocades from Italy or even further afield. Fashionable Italian silks of this period featured repeating patterns of roundels and animals, deriving from Ottoman silk-weaving centres in Bursa, and ultimately from Yuan Dynasty China via the Silk Road.

Cultural and costume historians agree that the mid-14th century marks the emergence of recognizable "fashion" in Europe. From this century onwards Western fashion changes at a pace quite unknown to other civilizations, whether ancient or contemporary. In most other cultures only major political changes, such as the Muslim conquest of India, produced radical changes in clothing, and in China, Japan, and the Ottoman Empire fashion changed only slightly over periods of several centuries.

In this period the draped garments and straight seams of previous centuries were replaced by curved seams and the beginnings of tailoring, which allowed clothing to more closely fit the human form, as did the use of lacing and buttons. A fashion for mi-parti or parti-coloured garments made of two contrasting fabrics, one on each side, arose for men in mid-century, and was especially popular at the English court. Sometimes just the hose would be different colours on each leg.

RENAISSANCE AND EARLY MODERN PERIOD

Renaissance Europe

Wool remained the most popular fabric for all classes, followed by linen and hemp. Wool fabrics were available in a wide range of qualities, from rough undyed cloth to fine, dense broadcloth with a velvety nap; high-value broadcloth was a backbone of the English economy and was exported throughout Europe. Wool fabrics were dyed in rich colours, notably reds, greens, golds, and blues.

Silk-weaving was well-established around the Mediterranean by the beginning of the 15th century, and figured silks, often silk velvets with silver-gilt wefts, are increasingly seen in Italian dress and in the dress of the wealthy throughout Europe. Stately floral designs featuring a pomegranate or artichoke motif had reached Europe from China in the previous century and became a dominant design in the Ottoman silk-producing cities of Istanbul and Bursa, and spread to silk weavers in Florence, Genoa, Venice, Valencia and Seville in this period.

As prosperity grew in the 15th century, the urban middle classes, including skilled workers, began to wear more complex clothes that followed, at a distance, the fashions set by the elites. National variations in clothing increased over the century.

Early Modern Europe

By the first half of the 16th century, the clothing of the Low Countries, German states, and Scandinavia had developed in a different direction than that of England, France, and Italy, although all absorbed the sobering and formal influence of Spanish dress after the mid-1520s..

Elaborate slashing was popular, especially in Germany. Black was increasingly worn for the most formal occasions. Bobbin lace arose from passementerie in the mid-16th century, probably in Flanders. This century also saw the rise of the ruff, which grew from a mere ruffle at the neckline of the shirt or chemise to immense cartwheel shapes. At their most extravagant, ruffs required wire supports and were made of fine Italian reticella, a cutwork linen lace. By the turn of the 17th century, a sharp distinction could be seen between the sober fashions favored by Protestants in England and the Netherlands, which still showed heavy Spanish influence, and the light, revealing fashions of the French and Italian courts.

The great flowering of needlelace occurred in this period. Geometric reticella deriving from cutwork was elaborated into true needlelace or punto in aria (called in England "point lace"), which reflected the scrolling floral designs popular for embroidery. Lacemaking centers were established in France to reduce the outflow of cash to Italy.

Dr. Wolf D. Fuhrig, "By the second half of the 17th century, Silesia had become an important economic pillar of the Habsburg monarchy, largely on the strength of its textile industry."

INDUSTRIAL REVOLUTION AND MODERN TIMES

During the industrial revolution, production was mechanised with machines powered by waterwheels and steam-engines. Sewing machines emerged in the nineteenth century.

Synthetic fibers such as nylon were invented during the twentieth century. Clothing and textile manufacture expanded as an industry so that such unions as the Amalgamated Clothing Workers of America and the Textile Workers Union of America formed early in the twentieth century. Later in the twentieth century, the industry had expanded to such a degree that such educational institutions as UC Davis established a Division of Textiles and Clothing, The University of Nebraska-Lincoln also created a Department of Textiles, Clothing and Design that offers a Masters of Arts in Textile History, and Iowa State University established a Department of Textiles and Clothing that featurs a History of costume collection, 1865-1948. Even high school libraries have collections on the history of clothing and textiles.

Alongside these developments were changes in the types and style of clothing worn by humans. During the 1960s, had a major influence on subsequent developments in the industry.

Textiles were not only made in factories. Before this that they were made in local and national markets. Dramatic change in transportation throughout the nation is one source that encouraged the use of factories. New advances such as steamboats, canals, and railroads lowered shipping costs which caused people to buy cheap goods that were produced in other places instead of more expensive goods that were produced locally. Between 1810 and 1840 the development of a national market prompted manufacturing which tripled the output's worth. This increase in production created a change in industrial methods, such as the use of factories instead of hand made woven materials that families usually made.

The vast majority of the people that worked in the factories were women. Women went to go work in textile factories because of some of the following reasons. Crowding at home was indeed a cause for them to leave and be on their own. The need to save for future marriage portions also motivated these women to decide to work in the millhouses. The work enabled them to see more of the world, to earn something in anticipation of marriage, and to ease the crowding within the home. They also did it to make money for family back home. The money they sent home was to help out with the trouble some of the farmers were having. They also worked in the millhouses because they could gain a sense of independence and growth as a personal goal.

USES OF TEXTILES

Textiles have an assortment of uses, the most common of which are for clothing and containers such as bags and baskets. In the household, they are used in carpeting, upholstered furnishings, window shades, towels, covering for tables, beds, and other flat surfaces, and in art. In the workplace, they are used in industrial and scientific processes such as filtering. Miscellaneous uses include flags, backpacks, tents, nets, cleaning devices such as handkerchiefs and rags, transportation devices such as balloons, kites, sails, and parachutes, in addition to strengthening in composite materials such as fibreglass and industrial geotextiles. Children can learn using textiles to make collages, sew, quilt, and make toys.

Textiles used for industrial purposes, and chosen for characteristics other than their appearance, are commonly referred to as *technical textiles.* Technical textiles include textile structures for automotive applications, medical textiles, geotextiles, agrotextiles, protective clothing (*e.g.* against heat and radiation for fire fighter clothing, against molten metals for welders, stab protection, and bullet proof vests). In all these applications stringent performance requirements must be met. Woven of threads coated with zinc oxide nanowires, laboratory fabric has been shown capable of "self-powering nanosystems" using vibrations created by everyday actions like wind or body movements.

Once a textile fabric has been manufactured, its potential for causing damage to the environment does not end. A fabric, no matter how high its quality or how well it is finished, coloured and generally enhanced, is of little or no use unless it is made into something. The possible products of this further processing include such widespread objects as clothing, upholstery, drapery, bedding and industrial goods. We normally tend to regard the primary (*i.e.* fabric production) and secondary (*i.e.* products made from fabrics) textile industries as separate ones, but the environment makes no such distinction; pollution is pollution no matter what its source may be.

One of the major problems facing the entire industry today is a lack of communication between these two different areas, since there is often a misinterpretation of the needs of the secondary sector by members of the first one, as well as a misunderstanding of the capabilities of the primary industry by the secondary one. Nature may benefit substantially by the development of a closer connection between the two, reducing the number of discarded materials that result from these failures to match needs, abilities and production.

TYPES OF USE

A distinction is proposed here between different types of use in a way not normally adopted. Those uses intended to protect the environment in some way will be separated from the others and treated. This is obviously not a simple matter, because there will be immense grey areas where the two cannot readily be distinguished. A geotextile barrier designed to prevent flooding of houses in an area inundated with water from a burst river bank, for instance, is clearly intended to keep the human inhabitants of the threatened region safe and alive. A geotextile used to contain polluted substances in a landfill site from leaching out into the water table is intended to protect the environment.

Yet, the flood barrier also prevents erosion of land by water flow, while the contained pollution is also prevented from poisoning human beings, so that both end results are met in each case. The distinction here (which is totally arbitrary and personal) is to determine what seems to be the more important criterion for use and to allocate the application to the section that makes most sense in that case.

NORMAL USES

Clothing

The most obvious normal use of textiles is in making clothing, a use that has been around for thousands of years. Products of the garment sector will include only clothing and accessories. As always, the machines needed to make them are complex, though usually much smaller in scale than those used in textile yarn and fabric manufacture. Cutting, sewing, pressing and other machines all need to be made and operated, at the now-familiar cost to the planet. There are frequently excess pieces of fabric that are cut away in making clothing to be discarded. It is, admittedly, possible to recycle these scraps, but large, complex machinery is needed to put them back into the production train.

Although such scraps may be too small anyway to make recycling worthwhile, one author describes yarns made with fibres recovered from the garment industry, noting that they avoid excess pollution by eliminating the need for fertilisers and dyes. Uses as socks, blankets and upholstery materials are mentioned specifically, a point that gives the key to this over-optimistic attitude. These articles differ from most textile end uses in that colour is not critical, unlike virtually all other applications. In more usual recycled goods, especially as they are almost invariably coloured, some process for removing all the dyestuffs is always needed. This involves chemical reactions in which bleaching agents or dye strippers are used to destroy the dye molecules or change their molecular structure to render them colourless. These substances are waste products that can harm the Earth once discarded into the environment.

The Fashion Industry

One of the most powerful driving forces influencing the manufacture of clothing is the fashion industry, which has increased considerably in importance in the centuries since the Industrial Revolution. Clothing is one of the easiest ways to display wealth conspicuously, so it is now accepted among people in many parts of the world (and particularly in the developed nations) that new outfits should be purchased at every season of the year.

Seasonal variations in climate are an added major incentive for these expenditures and, in practice, the resulting annual production of clothing may be regarded as too great to be justified as environmentally responsible. Few people actually wear more than one outfit at a time, which means that many garments are stored in environmentally destructive plastic bags for a great proportion of their useful life.

After the season ends, it is currently fashionable, in a kind of lip service to planetary well-being, to 'recycle' the articles via alternative outlets, such as second-hand clothing stores or factory shops selling designer clothes at reduced prices. Even this gesture is not entirely altruistic in terms of the Earth's welfare; the loss of an article of clothing invariably leaves a gap that must be filled by something new in the wardrobe of the person making this supreme sacrifice. The premises where the unwanted garments are resold also need to be built, heated, lit, staffed and so on. Unfortunately, the pressures of the fashion industry appear to be increasing rather than declining, as would be preferable for ecological welfare. The problem, as always, lies in commercial interest. Modern people, especially in the developed world, are accustomed to the need to follow a fashion cycle on an annual or even seasonal basis, so that the fashion industry (and those connected with it) has a vested interest in ensuring a regular turnover of garments.

It is inevitable that a large reduction in sales would lead to a major loss in employment, because it is not only the textile and clothing manufacturers who would be out of work, but also the designers, alteration staff, marketing specialists, transportation personnel, retailers, magazine editors or artists, and anyone else with an ancillary interest in the fashion world. The cumulative psychological effect on Society of the need of all these people to be kept in employment is difficult to resist, meaning that the relentless drive to be up-to-date will continue for as long as we (or our planetary home) are prepared to tolerate it.

Environmental Aspects

The renewal of the fashion requirement per se does not merely cause environmental problems in the debate regarding recycling versus disposal. As a consequence of our modern lifestyle and technology, dust and dirt are omnipresent, and soiling necessitates emergency procedures in the form of laundering or dry cleaning.

Both of these can cause environmental difficulties. Laundering involves exposing the textile article to the combined effects of water, heat, agitation and detergent. Often, an optical bleach or a fabric softener is added to the mix, and machine drying is frequently used. Both of these procedures give a much more luxurious feel than do an untreated wash and outdoor drying on a clothing line.

The energy needed to heat the water and operate the machinery, together with the chemicals discarded into the drain, are a source of harm, as also are the heat needed to operate and the polluted exhaust emitted from the dryer. The thermal load from both washing and drying can also affect the surrounding air or water, making it unsafe for other species to exist in the neighbourhood of the discard point unless precautions are taken to reduce the rise in temperature occurring on discharge.

Dry Cleaning

Dry cleaning also has ecological consequences. The machinery needs to be built and operated, while the solvents used for extracting the dirt are often toxic or carcinogenic. Even though strenuous efforts are made to contain them and recycle them for later use, there is still an inevitable loss because of entrapment in the fibres or escape into the air via the garment transport path. Jipp describes nontoxic stain removal using solvents without chlorocarbons, so that the chemicals can be regarded as being kind to both fibre and environment. Nevertheless, all of the reagents used in maintenance, even these more harmless ones, have to be manufactured, with expensive loading on the environment.

Maintenance Chemicals

The actual chemical substances used in maintenance can be examined in more detail to ascertain their potential for harm. As a consequence of our current fascination for cleanliness, garments are washed or cleaned far more often than is necessary for the sake of health or even hygiene. For this reason, the quantities of detergent, softeners, bleaching agents, dry cleaning solvents or other chemicals expelled are enormous.

Detergents contain alkalis and organic chemicals that act as pollutants. They also often contain phosphates, used as 'builders' to enhance the effectiveness of the washing action, which are known to encourage the growth of algae in large bodies of water. These can take over the oxygen available in the water, preventing other species (both animal and vegetable) from having access to it, As a result, fish can die and the balance of the local aquatic plants can be disturbed seriously, bringing about major changes in the locality and eventually resulting in a dead body of water where fish no longer exist and weeds choke the entire surface. Fabric softeners, usually quaternary ammonium compounds, are becoming more and more popular. These bring about water pollution. They are produced by relatively complex chemical reactions, with the usual environmental concerns and their disposal brings about harmful changes in the local water supply.

Household Textiles

In the home, textiles find widespread use, from household furnishings such as carpets, cushions and table linens, to towels, bath mats or dish cloths. David Rigby Associates give an overview of the various types of textile applications in home furnishing products, suggesting that the largest growth in the foreseeable future will occur in non-wovens or fibrefill categories, rather than in the more traditional woven or knitted fabrics. Two unusual applications in the household category have been reported in the literature; one author writes of the usefulness of wool tightly woven and attached to gypsum board (in the form of decorative panels) as a means of combating the relatively new hazard of sick building syndrome by absorbing the formaldehyde reputedly responsible for this problem. The second application consists of a system used to wrap houses during construction to maintain acceptable working comfort, for craftsmen operating on the outside walls of the house in adverse weather conditions, with enough moisture vapour permeability to ensure that breathing comfort can be retained.

Technical Textiles

Bhonde gives an overview of textiles used in applications beyond those of clothing or household purposes, while Legler examines technical textiles from the perspective of recent advances. Both authors more or less agree on the list of end uses, including geotextiles, agrotextiles, industrial products (including composite reinforcement), automotive, space, protective and medical ones. Not mentioned specifically in this list is the use of textiles in building, which may be extended to the subject of architectural fabrics. Their adoption, according to Hill, reduces building time significantly and saves natural resources (a valuable advantage from our point of view!), while providing long life and cost-effective, aesthetically pleasing appearance.

Swedberg describes applications of tent structures using PTFE-coated fibreglass with steel tubing and cables and, in a later paper, the use of air-inflated beams for supporting massive structures in high wind or loading conditions with potential applications in aerospace or nautical projects. Other interesting applications are the use of a composite of high-tensile glass with epoxy resin to make a reinforcement beam as protection against earthquake collapse and wrapping corroded bridge columns with carbon fibre to prevent further damage.

House Construction

House construction can also be aided indirectly by fibres, when used as reinforcement for concrete or in other aspects of building. In such applications, the environment is preserved by preventing premature failure of the material, allowing it to last longer. Peter recommends Kevlar for concrete reinforcement, because it imparts low density, non-catastrophic failure (and resistance to repeated impacts) to the concrete. In addition, because the fibres are electrically insulating, Kevlar permits building to take place more safely near high-voltage power lines.

Komatsu assess the durability of geotextiles in reinforcing asphalt concrete, noting that this process increases viscosity of the material and reduces the stress concentration of wheel loading. Locatelli reports the production of a stretchable roofing felt that can be installed and maintained without shrinkage when used in a layered system. A French company is marketing a woven structure that can be laid across soft, sandy or swampy ground, allowing vehicles to pass safely. In temporary houses, such as tents, fabrics are invariably used for their combination of light weight, easy folding ability, weather (especially wind) resistance and versatility.

Motor Vehicles

Fung provides a detailed description of the many applications of textiles, in the form of coated or laminated fabrics, in vehicles. He includes in his listing seat materials, headliner structures, other interior coverings, air bag components, convertible coverings, liners for bonnets or wheel arches, carpeting, noise control items and drive belts.

In view of current traffic hazard concerns involving motor vehicles, the air bags used to enhance safety in crashes have received particular attention. One writer recommends the use of nylon 4.6 in making them, because of its better thermal stability.

Barnes and Rawson, after devising a new test to assess the efficiency of coated fabrics for this end-use, find that nylon 6.6 is better than either nylon 6 or polyester in this regard. They also report that a silicone-based coating should be applied to the fabric to increase thermal resistance still further.

Gutlein *et al.* recommend a polyurethane bladder in a textile net for making side-impact bags, while yet another author would like to see needle-punched construction methods used to increase permeability in a controlled manner and describes conditions for achieving optimum gas transfer characteristics.

Medical Textiles

Medical uses of textiles should also not be ignored, as they constitute a major area of advance for the industry.

Industrial Applications

The widespread use of textiles in industrial applications, however, is not generally discussed in any detail. Various authors write articles dealing with the many valuable attributes of technical fabrics, including those intended for space activity, automotive filtration, airbags, composites, marine uses, laminating, seismic protection and resistance to harmful agents.

Applications range widely in type and some consideration of representatives of all the varieties of application should be included. They may be divided into those used directly by industry, those used in leisure or sport activities and those in which heavy duty use is needed in everyday life. Textiles used directly in industry include such diverse items as tarpaulins, filters, acoustic or thermal insulation and protective garments for workers.

Other Uses

Three specific types of more interesting use can be identified. First, transport via heavy lorries is rendered less perilous (and has a lower risk of polluting the road or causing accidents by losing part of the material being shipped) by the use of a new kind of tarpaulin, knitted from stretchable textured polyester, that conforms to the shape of a load, reducing flapping as the vehicle moves.

Second, filters appear to be popular subjects for research, an occurrence caused at least in part by the need for a cleaner environment. Third, leisure or sport applications are rapidly increasing and include items such as tentage, sleeping bags, boat sails, mooring ropes or other marine cordage, aircraft skins, parachute fabrics, climbing ropes, bungee cords, balloon fabrics or guy ropes. Little attention appears to have been paid to most of these applications, although there are some papers worth mentioning in the literature.

Finally, one or two less obvious uses may be mentioned. Newberry reviews new developments in the use of composites in corrosion-resistant applications such as piping, lamp poles, and tanks to contain hot brine or for use in deionised brine service. Risks of harm from mobile phones, reportedly causing brain damage from electromagnetic radiation, are supposedly counteracted by the use of a flexible metal-coated fibre, preferably an acrylic one coated with copper or nickel. Plates used to press boards of plywood, or those for printed circuits, lined with textile cushions designed for high-temperature use, are reported, while Marsh discusses the use of fibre-reinforced plastics for improving under-sea oil extraction; the smooth surface, chemically inert nature and low weight mean that lower losses from corrosion and oil escape will occur, so providing an ecological benefit. Kotliar uses textile and carpet waste to produce a low-cost, wood-like material by incorporating them in a high-modulus phenol/formaldehyde matrix.

Other heavy-duty uses in everyday life include tyre cords, seat belt webbing, vehicle upholstery, carpeting, wall hangings, conveyor belts, drive belts, tow ropes, isolation suits, protective clothing, space suits, architectural fabrics and geotextiles. In addition, cleaning cloths, surgical clothing or drapes, bandages or other dressings should also be added to the list.

OBJECTS OF PRESERVING TEXTILE

There are several techniques that can be used to avoid exposing textiles to the above deterioration agents. Many ways of repairing and cleaning textiles also exist; however, aside from careful vacuuming, these techniques are not recommended for those who are not trained in textile conservation. Many conservation methods have the potential to damage the textile further, and even professional conservators consider many factors before making the decisions on how to treat and repair a textile object. Wet cleaning is a widely used cleaning method, but it is extremely hazardous when done improperly. Loss of dyes or paints, the removal of water-soluble sizes, dimensional changes, and the production of a kind of "textile mud" when highly degraded textiles are exposed to water are among the possible results of wet cleaning.

Another situation that requires caution and expert advice is when folded, aged textiles are discovered. If a textile object is found folded, it is best to contact a professional conservator for advice before attempting to unfold it. Even though the object may appear to be in stable condition, deep creases can cause damage over time causing the fabric to split when it is unfolded. Professional treatments such as humidifying may be necessary before the textile can be flattened.

Non-professionals should focus on preservation methods that carry as little risk as possible for causing further damage. Fortunately, adequate preservation has become increasingly easy for the general public and entities with smaller budgets due to the availability of relatively affordable archival products over the Internet and elsewhere. A short list of companies with Internet sites that offer archival materials for sale. This list is provided only to aid in the search for materials and does not entail promotion of the establishments listed.

VACUUMING

In some cases, a textile may have surface dirt and dust that can be removed by vacuuming if the textile is not excessively brittle. The advice of a professional is recommended before this is performed to ensure that the textile is stable enough to endure this process. Dust removal is an important first step because layers of dust can carry aggressive corrosive substances such as sulphur dioxide, nitorgen monoxide, or various oily substances. Regular surface cleaning by vacuuming is necessary for any textile that is on display as well as for textiles with large amounts of surface soiling. A monofilament screening attached to an open frame should be placed screen side down on top of the area to be vacuumed to prevent fibre loss. The screen should be heavy enough to lie firmly on the textile during vacuuming, and the edges of the screen should be taped to prevent snagging of the textile. The hand held nozzle attachment of a vacuum with a low suction setting or a low powered hand vacuum can then be used directly on top of the screen to remove dust and surface dirt. Any textile object that is at a point where small pieces are breaking off should not be vacuumed at all even under a screen because textile loss may occur.

It is also possible to surface clean a textile using a hand held nozzle or vacuum with the opening covered with fine net. This method is not usually recommended, however, because the vacuuming apparatus must be held slightly above the textile and cannot touch the textile directly without risking damage to the object. Much more caution is required for this method than for the screen method. Brushing is not recommended because it only moves dust around instead of removing it and has the potential to cause damage even when done lightly.

TECHNICAL OR INDUSTRIAL TEXTILES

For many years, the term 'industrial textiles' was widely used to encompass all textile products other than those intended for apparel, household and furnishing end-uses. It is a description still more widely favoured in the USA than in Europe and elsewhere.

Technical Textiles Market

This usage has seemed increasingly inappropriate in the face of developing applications of textiles for medical, hygiene, sporting, transportation, construction, agricultural and many other clearly non-industrial purposes. Industrial textiles are now more often viewed as a subgroup of a wider category of technical textiles, referring specifically to those textile products used in the course of manufacturing operations (such as filters, machine clothing, conveyor belts, abrasive substrates etc.) or which are incorporated into other industrial products (such as electrical components and cables, flexible seals and diaphragms, or acoustic and thermal insulation for domestic and industrial appliances).

If this revised definition of industrial textiles is still far from satisfactory, then the problems of finding a coherent and universally acceptable description and classification of the scope of technical textiles are even greater. Several schemes have been proposed. For example, the leading international trade exhibition for technical textiles, Techtextil (organised biennially since the late

1980s by Messe Frankfurt in Germany and also in Osaka, Japan), defines 12 main application areas (of which textiles for industrial applications represent only one group):

- *Agrotech*: Agriculture, aquaculture, horticulture and forestry
- *Buildtech*: Building and construction
- *Clothtech*: Technical components of footwear and clothing
- *Geotech*: Geotextiles and civil engineering
- *Hometech*: Technical components of furniture, household textiles and floorcoverings
- *Indutech*: Filtration, conveying, cleaning and other industrial uses
- *Medtech*: Hygiene and medical
- *Mobiltech*: Automobiles, shipping, railways and aerospace
- *Oekotech*: Environmental protection
- *Packtech*: Packaging
- *Protech*: Personal and property protection
- *Sporttech*: Sport and leisure.

The search for an all embracing term to describe these textiles is not confined to the words ‘technical’ and ‘industrial’. Terms such as performance textiles, functional textiles, engineered textiles and high-tech textiles are also all used in various contexts, sometimes with a relatively specific meaning (performance textiles are frequently used to describe the fabrics used in activity clothing), but more often with little or no precise significance.

Inconsistent Statistical Reporting

To add to this complexity, different geographical regions and countries tend to adopt rather different viewpoints and definitions with regard to all of the above. A widely quoted misconception that technical textiles in Japan account for over 40% of all textile output or nearly twice the level in Western Europe can largely be put down to the different statistical bases employed. In Europe, the most authoritative source of fibre consumption (and therefore textile output) data is CIRFS (Comité International de la Rayonne et des Fibres Synthétiques), the European artificial fibre producers association.

However, CIRFS’ reported statistics (at least until recently) have specifically excluded tape and film yarns (a significant proportion of all polyolefin textiles), coarser monofilaments and all glass products (as well as natural fibres such as jute, flax, sisal, etc.). The merger of CIRFS and EATP, the European Polyolefin Textiles Association, should go some way towards resolving this anomaly.

The Japanese ‘Chemical’ Fibres Manufacturers Association, JCFA, at the other extreme, includes all these products, including natural fibres, within its definition of technical/industrial textiles while the Fiber Statistics Bureau in the USA includes polyolefin tape and monofilament yarns but excludes glass.

The relative usage of the main technical fibres and yarns on a more consistent basis. In this new light, Japan still retains a leading position worldwide in terms of the proportion of its total textile manufacturing output devoted to technical textiles. However, this is largely a reflection of the importance of its automotive manufacturing industry (a key user of technical textiles) combined with the relatively smaller size of its apparel and furnishing textile sectors (especially floor coverings). The USA apparently accounts for the lowest proportion of technical textile output of the three major industrial regions but still produces and consumes the largest quantity per capita, especially when all glass textile and technical fibre uses are included.

OPERATING AT THE BOUNDARIES OF TEXTILES

If the adjective 'technical' is difficult to define with any precision, then so too is the scope of the term textiles. The principal materials, processes and products which are commonly regarded as falling within the scope of technical textiles manufacturing. However, there remain many grey areas. For example, the manufacture and processing of metallic wires into products such as cables, woven or knitted screens and meshes, and reinforcing carcasses for tyres are not generally regarded as lying within the scope of the textile industry. This is despite the fact that many of the techniques employed and the final products obtained are closely related to conventional textile fibre equivalents.

Within the composites industry, woven, knitted, braided, nonwoven and wound yarn reinforcements made from glass, carbon fibre and organic polymer materials such as aramids are all now widely accepted as being technical textile products. On the other hand, more loosely structured reinforcing materials such as chopped strand mat, milled glass and pulped organic fibres are often excluded.

The nonwovens industry has developed from several different technology directions, including paper manufacturing. The current definition of a nonwoven promulgated, for example, under the International Standards Organization standard ISO 9092 acknowledges a number of borderline areas, including wet-laid products and extruded meshes and nets. Likewise, distinctions between textile fibres and filaments, slit or fibrillated films, monofilaments and extruded plastics inevitably boil down to some fairly arbitrary and artificial criteria.

Diameter or width is often used as the defining characteristic, irrespective of the technologies used or the end-uses served. Many of the definitions and categories embodied within existing industry statistics reflect historical divisions of the main manufacturing sectors rather than a functional or market-based view of the products involved.

Polymer membranes, composite materials and extruded grids and nets are other products which challenge traditional notions of the scope of technical textile materials, processes and products. Increasingly, technical textiles are likely to find their place within a broader industry and market for 'flexible engineering materials'. A number of companies and groups have already adopted this outlook and operate across the boundaries of traditional industry sectors, focusing a range of materials, process technologies and product capabilities upon specific functions and markets such as filtration and health care.

DEVELOPMENT OF TECHNICAL TEXTILES

Although the development of technical and industrial applications for textiles can be traced back many years, a number of more recent milestones have marked the emergence of technical textiles as we know them today. Very largely, these have centred upon new materials, new processes and new applications.

Viscose Rayon

The first commercially available synthetic fibre, viscose rayon, was developed around 1910 and by the 1920s had made its mark as reinforcement material for tyres and, subsequently, other mechanical rubber goods such as drive belts, conveyors and hoses. Its relatively high uniformity, tenacity and modulus (at least when kept dry within a rubber casing), combined with good temperature resistance, proved ideal for the fast emerging automotive and industrial equipment markets. At a much later stage of its lifecycle, other properties of viscose such as its good absorbency and suitability for processing by paper industry-type wet laying techniques contributed to its role as one of the earliest and most successful fibres used for nonwoven processing, especially in disposable cleaning and hygiene end-uses.

Polyamide and Polyester

Polyamide (nylon) fibre, first introduced in 1939, provided high strength and abrasion resistance, good elasticity and uniformity as well as resistance to moisture. Its excellent energy absorbing properties proved invaluable in a range of end-uses from climbing ropes to parachute fabrics and spinnaker sails. Polyamide-reinforced tyres are still used much more extensively in developing countries where the quality of road surfaces has traditionally been poor as well as in the emerging market for off-road vehicles worldwide. This contrasts to Western Europe where average road speeds are much greater and the heat-resistant properties of viscose are still valued.

From the 1950s onwards, the huge growth in world production of polyester, initially for apparel and household textile applications, provided the incentive and economies of scale needed to develop and engineer this fibre as a lower cost alternative to both viscose and polyamide in an increasing range of technical applications. Nowhere is this more true than Japan and the developing industrial economies of Asia, including China, where production capacities for both polyester staple and filament yarn are extremely high and there is an urgent search for new applications. Some high volume applications for technical textiles which would typically use poly-olefins in western Europe and North America such as geotextiles, carpet backing and coverstock are more likely to use polyester in Asia largely because of the greater availability and better economics of fibre supplies in those regions. At a slightly less obvious level, differences in the polyamide supply situation -Western Europe and North America are more strongly oriented towards nylon 66 while Asia and Eastern Europe produce predominantly nylon 6 - are reflected in different manufacturing practices, product types and technical applications for this fibre.

Yet another example is the production and use of Vinylon (PVA, polyvinyl alcohol) fibres in Japan, where they were developed for a variety of industrial and technical applications at a time when that country lacked other raw materials and fibre production capabilities. Use of this fibre for technical textiles is almost non-existent in the West.

Polyolefins

The development of polyolefin (mostly polypropylene but also some polyethylene) fibres as well as tape and film yarns in the 1960s was another milestone in the development of technical textiles. The low cost and easy processability of this fibre, combined with its low density and good abrasion and moisture-resistant properties, have allowed its rapid introduction into a range of applications such as sacks, bags and packaging, carpet backings and furniture linings as well as ropes and netting. Many of these markets were directly taken over from jute and similar fibres but newer end-uses have also been developed, including artificial sports surfaces.

Properties of the polyolefins such as their poor temperature resistance and complete hydrophobicity have been turned to advantage in nonwovens. Initially used in conjunction with viscose to permit thermal bonding, polypropylene has now benefited from a growing appreciation of the important role that moisture wicking (as opposed to absorption) can play in hygiene applications such as coverstock for diapers (nappies). Finally, the relatively low extrusion temperatures of the polyolefins have proved ideally suited to the fast developing technologies of spin laying (spun bonding and melt blowing).

The development of the polypropylene industry was initially focused on European and North American markets. However, it is undergoing a major expansion worldwide as new investment in polymer capacity offers more favourable economics to new geographical markets.

High Performance Fibres

The above 'conventional' fibre types, both chemical and natural, still account for over 95% of all organic fibre technical textiles in use (*i.e.* excluding glass, mineral and metal fibres). Many of them have been modified and tailored to highly specific end-uses by adjustment of their tenacity, length, decitex, surface profile, finish and even by their combination into hybrid and bicomponent products. However, it is the emergence of the so-called high performance fibres since the early 1980s that has provided some of the most significant and dramatic impulses to the evolution of technical textiles. First and foremost of these are the aramids, both the highly temperature-resistant *meta*-aramids (widely used in protective clothing and similar applications) and the high strength and modulus *para*-aramids (used in a host of applications ranging from bulletproof vests to reinforcement of tyres, hoses, friction materials, ropes and advanced composites).

From their commercial introduction in the 1970s, world demand for *p*-aramids is expected to reach almost 40000 tonnes per annum by 2000 while for *m*-aramids, consumption will be around 17-18000 tonnes. While not huge in overall terms (representing less than 0.5% of total world technical fibre and yarn usage in volume terms but closer to 3-4% in value), the aramids represent a particularly important milestone in the development of the technical textiles industry.

Partly practical and partly symbolic, the introduction of the aramids not only led to the injection of large amounts of technical and market support into the industry and for users by leading fibre manufacturers such as DuPont and Akzo, but also concentrated the minds of many developers of new products upon the possibilities (and practicalities) of using similar new generation materials. The early success of the aramids was a welcome contrast to the development of carbon fibres, which have been commercially available since the 1960s but largely constrained by their high material and processing costs to selected high value markets, particularly aerospace applications. Total world demand for carbon fibres was still only some 8-9000 tonnes per annum as recently as 1995. In fact, their market actually shrank in the early 1990s owing to cutbacks in military spending.

At long last, carbon fibres appear to be emerging from the doldrums, with the appearance not only of important new civil aerospace markets but also of high technology sporting goods and industrial applications such as wind generator turbine blades and reinforced fuel tanks. As new manufacturing methods and greater economies of scale start to bring prices down, the feasibility of even larger scale applications such as the reinforcement of buildings and structures in earthquake zones becomes more attractive. Currently, (2000), consumption is considered to be over 13 000 tonnes per annum, rising to almost 19 000 tonnes by the year 2005.

The introduction of other high performance fibres proliferated, particularly during the late 1980s, and in the wake of the aramids. These included a range of heat and flameproof materials suitable for protective clothing and similar applications (such as phenolic fibres and PBI, polybenzimidazole), ultra-strong high modulus polyethylene (HMPE) for ballistic protection and rope manufacture, and chemically stable polymers such as polytetrafluoroethylene (PTFE), polyphenylene sulphide (PPS) and polyethyletherketone (PEEK) for use in filtration and other chemically aggressive environments. Individually, none of these other fibres has yet achieved volume sales anywhere near those of the aramids (or even carbon fibres). Indeed, the output of some speciality fibres can still be measured in tens of tonnes per year rather than hundreds or thousands. The widespread industrial recession of the early 1990s caused many fibre manufacturers to review their development strategies and to focus upon narrower ranges of products and markets.

Glass and Ceramics

Glass has, for many years, been one of the most underrated technical fibres. Used for many years as a cheap insulating material as well as a reinforcement for relatively low performance plastics (fibre glass) and (especially in the USA) roofing materials, glass is increasingly being recognised as a sophisticated engineering material with excellent fire and heat-resistant properties. It is now widely used in a variety of higher performance composite applications, including sealing materials and rubber reinforcement, as well as filtration, protective clothing and packaging.

The potential adoption of high volume glass-reinforced composite manufacturing techniques by the automotive industry as a replacement for metal body parts and components, as well as by manufacturing industry in general for all sorts of industrial and domestic equipment, promises major new markets. Total world con- sumption of 'textile' glass in technical applications was some 2.3 million tonnes per annum in 1995 and is considered likely to be over 2.9 million tonnes at 2000, representing over 20% of all technical fibre consumption. Various higher performance ceramic fibres have been developed but are restricted to relatively specialised applications by their high cost and limited mechanical properties.

DEVELOPMENTS IN FIBRE MATERIALS - NATURAL FIBRES

Until early in the 20th century, the major fibres available for technical and industrial use were cotton and various coarser vegetable fibres such as flax, jute and sisal. They were typically used to manufacture heavy canvas-type products, ropes and twines, and were characterised by relatively heavy weight, limited resistance to water and microbial/fungal attack as well as poor flame retardancy. Some of the present day regional patterns of technical textiles manufacturing were established even then, for example Dundee, on the east coast of Scotland and located at the centre (then) of an important flax growing area as well as being a whaling port. Following the discovery that whale oil could be used to lubricate the spinning of the relatively coarse jute fibres then becoming available from the Indian subcontinent, jute fabrics were widely used for sacking, furniture and carpet manufacturing, roofing felts, linoleum flooring, twine and a host of other applications.

Although its jute industry was to decline dramatically from a peak at around 1900 owing to competition from other materials as well as from cheaper imports, Dundee and the surrounding industry subsequently become a nucleus for development of the UK polypropylene industry in the 1960s. The then newly available polymer proved not only to be an ideal technical substitute for the natural product but was also much more consistent in terms of its supply and price.

Traditional end-uses for sisal were similarly rapidly substituted throughout the established rope, twine and net making centres of Europe and America. Wool proved far less versatile and economic for most industrial applications although it is still valued for its insulating and flame retardency properties and finds use in several high temperature and protective clothing applications.

Silk is an even more exotic fibre, rarely used in technical applications other than for highly specialised uses such as surgical suture thread. However, the traces of the early silk industry are still to be seen in the present day location of centres for technical filament weaving such as the Lyons area of France. The traditional silk industry has also contributed to the development of technical textiles in Asia, especially in Japan.

Textile Processes

The wide range of processes employed in the manufacture of technical textiles. Apart from the use of plaiting and knotting for the manufacture of ropes and nets, weaving was, for many years, the pre-eminent technology employed in the manufacture of 'industrial' textiles. In terms of the total weight of textiles produced, weaving still plays a leading role and developments such as three-dimensional and crimpless weaving have opened up many new product and end-use possibilities.

However, the historical progress of technical textiles has seen the advance of alternative textile forming technologies, most prominently the broad family of non-woven techniques but also warp and weft knitting, stitchbonding and modern braiding methods. The use of loose fibres with sophisticated cross-sectional profiles for insulation, protection and fibrefill applications is another important growth area. Fibres, yarns and textiles of all types also provide the starting point for a diverse and fast expanding range of composite reinforcement and forming technologies.

According to a major study of the world technical textiles industry and its markets projected to 2005, nonwovens are set to overtake weaving (in terms of the total weight of textiles produced) by around 2002/2003. In area terms, nonwovens already far exceed woven and other fabric forming methods because of their lower average weight per unit area. On the other hand, woven and other yarn-based fabrics will remain in the lead in value terms, at least for the foreseeable future. There is, therefore, something for every section of the textile industry in the future of technical textiles. Most product areas will see more rapid growth in value than in volume as technical textiles become increasingly sophisticated and employ more specialised and higher value raw materials. On the other hand, the total value of yarns and fibres and of all technical textile products will grow slightly less fast than their volume because of a changing mix of materials and technologies, especially reflecting the growth of nonwovens.

TECHNOLOGICAL CHANGE IN THE TEXTILE INDUSTRY

Technological changes in the textile industry during the last forty years can be broadly divided into three phases. High-speed spinning frames and looms, with reduced vibration levels, were developed in the 1950s and early 1960s. The most radical alterations in the core technology of spinning and weaving came in the late 1960s and during the 1970s, with the introduction of rotor spinning and shuttleless looms.

This was the period when new technology was sought to increase productivity and thereby to combat the cost-based competitiveness of lower wage producers from Asian countries. From the late 1970s onwards, the changes in the textile industry of the developed countries have been characterized by the introduction of microelectronics based technology and the automation of industrial processes.

The new technologies have allowed the various production stages to become one continuous process of interrelated activities, resulting in higher quality and flexibility and thus a faster response to changing market conditions. The full exploitation of these technological improvements is dependent on further complementary changes in organization and management. Du©ring the 1990s, rapid response to market conditions is likely to become an even more important force behind technological innovation and is likely to focus on increases in speed and flexibility rather than on cost-cutting.

Investment costs associated with technology improvements in the textile sector have risen considerably. About US $1 billion per year has been invested in the European Union, and twice as much in the USA, throughout the late 1980s and early 1990s. These levels could double in the coming years. The impact of automation on employment has also been significant. During the last 15 years, employment in the textile and clothing industry in the entire EU area has declined by 40 per cent, and the forecast for the 1990s is a loss of 700,000 to I million jobs in the textile sector alone. The occupational structure has also changed, with the proportion of operators and unskilled labourers decreasing while the share of technical and management staff increases. The new technologies require specialized skills in textile engineering, maintenance, design, computer science, and marketing.

TECHNOLOGICAL CHANGE IN THE GARMENT INDUSTRY

The pace of technological innovation in the garment industry was slow up to the beginning of the 1980s. The main innovations before 1980 were in sewing technology, with faster and more durable machines and the development of attachments for specialized tasks, which later resulted in the emergence of task-dedicated machines. In the pre-assembly phase the introduction of fast, automated cutters in the late 1970s replaced the operator-driven hydraulic die cutting machine.

The major technological changes occurred in the 1980s, when microelectronics penetrated all stages of garment production to some degree. The most significant innovations took place in the pre-assembly stage, where computer-aided design (CAD), computer numerical control (CNC) cutting systems, and computer-aided manufacturing (CAM) led to material and labour savings. In the assembly stage, which accounts for 80 per cent of the manufacturing value added (MVA) and of the workforce in the industry, technological change has so far been relatively modest.

The main improvements have been microelectronic control units which are attached to the standard industrial sewing machine to handle the more complex tasks. These can either be used to speed up production on task dedicated machines or to increase the flexibility of multi-purpose machines. The major technological innovations of the 1990s are likely to build on these developments. The introduction of computer-based technology has increased the cost of clothing machinery enormously, yet nearly sixty advanced CAD systems are already in Hong Kong, the Republic of Korea and Thailand alone, with over 60 per cent of these having been installed in the last three years.

Organizational Changes in Textiles and Clothing

Organizational changes complement the technological drive towards greater market responsiveness. These changes affect not only production itself, but also the relationship between suppliers and buyers. The emphasis in the organization of production is shifting to a multi-skilled workforce, closer management-workforce cooperation, and increasing reliance on externally supplied industrial services.

There is also a tendency towards vertical integration within the textile and garment industry. In the field of buyer-supplier relations, textile producers, garment makers and retailers are working together in design teams. Price is no longer the only important determinant of sourcing: elements such as quality, timely delivery and technical cooperation between buyer and supplier play an increasingly important role. The buyer-supplier relationship is also intensified by the reduced lead time and shortened production runs. This means that close proximity to the markets and availability of efficient international telecommunication and transport networks have become an important competitive factor for the producer countries.

EMPIRICAL EVIDENCE FROM COUNTRY CASE STUDIES

The following characteristics of the textile and garment sector were identified in the case studies of Bangladesh, Thailand and Indonesia.

Resource Endowments - The Textile Industry

The presence of natural resources - mainly cotton, but in the case of Thailand also silk - was an important initial factor in the emergence of the textile industry in Bangladesh (when it was a part of India), Indonesia and Thailand. In all three countries the development of modern textile manufacturing, especially for export, was based on extensive use of abundant low-cost labour. This factor has made the industry an important source of employment and secured it a competitive advantage in export markets. Bangladesh and Indonesia have retained this relative advantage, but this is no longer the case for Thailand.

In Thailand, local raw material production has not grown sufficiently, and cotton yarn is now imported on a large scale, as well as man-made fibres. The quality of domestic raw silk is also hampering the modernization of handloom production, but joint R&D efforts are being undertaken by the Government and private business to introduce new varieties of raw silk suited to more advanced technology.

In Indonesia, local production of man-made fibres has been facilitated by the local availability of oil and timber. While the production of man-made fibres has developed sufficiently to cover most of the textile industry's requirements, most of the raw cotton is imported.

The Garments Industry

The modem garments industry developed much later than the textile industry in all three countries. At the initial stage of the import substitution period, the industry benefited from upstream linkages with the textile sector in Thailand and Indonesia. In Bangladesh, with no substantial local demand for manufactured garments, the industry was export-driven from the start.

Today, Bangladesh and Thailand rely increasingly on imported textiles, because local textile producers are no longer capable of providing the quantity, quality and variety of cloth required for export garments. In Indonesia, local fabric still satisfies almost 80 per cent of the industry's demand. In all three countries the competitive strength of the garment industry has been built upon a cheap labour force, predominantly women.

The Sub-sector's Role in Manufacturing

In Bangladesh, the textile branch accounts for some 35 per cent of MVA, down from over 50 per cent in the early 1970s. In Thailand and Indonesia (which are further along the path to a modern, diversified manufacturing sector) the textile sector accounted for 11 per cent and 10 per cent of MVA, respectively, at the end of the 1980s. The industry is still one of the largest employers of manufacturing labour in Bangladesh but has been surpassed by other industrial branches in Thailand and Indonesia. Its significance as an export earner has declined in all three countries. In Thailand and Indonesia this can largely be explained by the diversion of output to the local market.

In Bangladesh, textiles continue to be produced largely by traditional cottage industries (the handloom sector) although the pre-liberation period in the early 1960s saw a rapid expansion of integrated mills. Small-scale industries also play a very important role in Indonesia and Thailand (silk production).

Indonesian microenterprises have made more technical progress than those in Bangladesh and Thailand. In Thailand, however, improvements in the organization and management of subcontracting have increased the number of home-based silk producers and contributed to the growth of the silk industry. Public ownership of large-scale enterprises is more pronounced in Bangladesh than in Thailand and Indonesia, where local and foreign investment (private or joint ventures) dominate.

The MVA share of the garment industry is quite modest in all three countries: I per cent in Bangladesh, 7 per cent in Thailand, and 1.8 per cent in Indonesia. But the trend is still upwards, and the branch is one of the major contributors to foreign exchange earnings in spite of its low MVA share, especially in Bangladesh. Cheap, mass-produced clothes remain the main export item. In the Indonesian and particularly the Thai garment industries, there is now a trend towards quality products based on local design and diversification. In all three countries, small and medium-size enterprises predominate, often subcontracting for foreign firms which are responsible for high value added, industry-related, services such as design, distribution and marketing.

TECHNOLOGICAL CAPABILITIES IN TEXTILES

Overall, technology levels in the textile industry in the three countries have not changed much since the 1960s. Technological stagnation is most evident in Bangladesh. There is as yet no sign of automation in the large-scale mills, and traditional handloom weaving predominates in the informal sector, although semi-automatic Chittaranjan looms have been introduced on a modest scale.

In Thailand and Indonesia, advanced technologies are now being introduced, especially in foreign-owned mills producing synthetic textiles. But these technologies cover only isolated aspects of production, such as spinning. There are as yet no integrated modem mills. Thai producers have partly compensated for the lack of technological progress by better organization and maintenance. Efforts are also being made to modernize the traditional Thai silk industry. In Indonesia, power looms have become common in cottage industries. In all three countries, there is a traditional female presence in the textile industry, although in Bangladesh women were traditionally confined to yarn processing and excluded from weaving activities. This explains the unusually low share of women in the textile sector in Bangladesh, about 10 per cent compared to about 80 per cent in Thailand. No figures are available for Indonesia, but the rate can be estimated at between 40 and 60 per cent.

Small enterprises and subcontracting are important for female employment in all three countries. The introduction of a semi-automated loom in the traditional handloom sector, and the formation of female cooperatives, have created new employment for women weavers in Bangladesh. But in the 'modern' sector, outdated technology and a heavy reliance on physical strength are real obstacles to increasing female participation. The prevalence of traditional values and strong competition from men in the saturated labour market makes it even harder for women to increase their share.

In Thailand, it is too early to judge the effect of the recent modernization of the traditional silk industry on overall female employment in this sector. It is, however, possible that some of the newly-created opportunities for female factory employment, mostly for young and unmarried women, could be at the expense of women working as subcontractors from their homes.

In Indonesia, the available information suggests a decrease in employment opportunities in the traditional weaving sector and in small enterprises as the result of modernization and competition from the modern, large-scale sector. The growth of the modern sector seems to have compensated for this loss so far, but this is unlikely to continue.

TECHNOLOGY INDUSTRY OF TEXTILE

A high technology industry will employ fewer employees, but these will be more highly educated. The postulation for management and technology trained people will increase, but clerical jobs will decrease. The causes for the decline in employment are: import penetration, higher productivity with new technology, and the death of appropriate marketing aggressiveness of the manufacturers. These factors influence the different industry segments in various degrees. In the aggregate, rising imports influence the growth of the industry but not employment.

TEXTILE AND APPAREL INDUSTRY PRODUCTION

The textile and apparel industry production growth is still at 2%. Saving jobs is not a useful argument to keep the textile and apparel industry going. Textile and apparel industry segments seen to have completely different structure and performance profiles. For some industry segments, prospects are good, for others, prospects are bad. Employment, competitiveness, productivity and profitability are the critical variables. Prospects are good for weaving industries, non-woven industries, and the tufting industry.

Prospects are common for the throwing, winding, and thread industries and the knitted outerwear industry. Lastly, prospects are bad for the circular and warp knit industries, the hosiery and knit underwear industry, and the apparel industry, in general. The following are the reasons for good prospects of certain industries. The weaving industries invest heavily in new technology which increases productivity and profitability.

The technology is chiefly foreign development. The non-wovens industry is clearly in the early stages of the industry life cycle with heavy exports and increasing employment, whereas the tufting industry is fragmenting due to negative returns to scale. This implies that larger firms do not enjoy higher productivities due to their large size. This has, however, not led to decreases in employment. Exports are extensive.

Industries with average prospects are the knitted outerwear, throwing, winding, and thread industries. The knitted outerwear industry has high positive returns to scale in the productivity sense. This reasons consolidation, but with no effect on employment. The throwing, winding and thread industries substitute lower cost capital for high cost labour with increasing automation. There will be further investments in new technology in chiefly large scale plants. The industry is competitive. One can expect decreasing employment due to technologies. Mainly bad prospects face the apparel and knitting-related industries. The hosiery and knitted outerwear industries are disinvesting due to imports and not technology. In the circular and warp knit industries, both technology and imports reduce employment.

For instance, highly flexible and innovative marketing of sweaters by foreign firms has gained a large share of the sweater market, but the foreign firms have no technological advantage over firms. Besides, technological advances do not improve the financial performance. The apparel industries are depressed by large imports. New technology reduces the labour content, but does not always improve productivity. In general, the textile industries can benefit from investments in new technology which is chiefly produced in Europe and Japan to improve productivity and profitability with a reduction of labour content.

The apparel industries can reduce labour content to a certain point. However, this has not, so far, importantly improved productivity. In the aggregate, both industries will face large decreases in labour content. Some apparel industries must advantage from their potential in marketing by getting international marketing expertise. There are potential growth areas in the textile and apparel industry. Productivity in the textile industry has increased faster than the industry average over the past twenty years. The potential for further increases in productivity is high, but increased productivity does not essentially yield short run financial improvements.

Long-term planning is required, and a particular independence from the financial markets is beneficial. Privatization of textile companies is very common. Productivity increases are mainly geared towards reducing the labour content in production. Higher productivity machinery is chiefly supplied by overseas manufacturers. The lost its competitiveness in the development of textile machinery. Further developments include system-wide improvements in materials handling and more efficient work flows rather than just invention of new gadgets. Productivity increases in the apparel industry are slow. They all emphasize the reduction of labour content. The sewing process is the limitation to higher productivity increases.

There have been no revolutionary improvements in systems to join fabrics, including sewing systems. System wide improvements promise higher returns than "futuristic" technologies. Improved management and control systems can drastically improve productivity of the textile and apparel industry, overall. Inventory control, the optimization and speeding up of order-to-delivery time are deasive. The time from order placement to delivery can be decreased by improved cooperation of tow difficult industry segments.

Fragmentation of the Industry

The fragmentation of the industry must be decreased in the sense that firms integrate vertically to achieve economies of scale in productivity and flexibility with fast response to the market. Vertical integration is not necessary in the sense of finance and ownership. However, this means that hostility among the industry segments disappears. Most system-wide improvements require coordination between the industry parts without which the full impact of these technologies will not be realised. Productivity in the textile industry is the most dominant factor in the competition among producers. The entire productivity is somewhat less important in the apparel industry against foreign competitors since there are other factors like exchange rates and high labour costs which dominate.

Productivity advantages over competitors can only be reached with regular upgrading of machinery. Since the technologies in these industries, particularly in the weaving and spinning sectors, change rapidly, the machinery becomes obsolete in short time periods. The capital needs, for example, for approximately five year cycles in the spinning industry, are enormous. Only financially strong industries are able to upgrade their production technologies without assistance.

There are three important technology related factors which keep the textile and apparel industries from being competitive and dominant in the international markets.

1. The textile and apparel industries receive the same technologies as do competitors in the world markets. The textile and apparel industries have no technological benefit over their foreign competitors.
2. The textile and apparel industries are dependent on foreign suppliers for machinery.
3. The textile machinery industry is fastly declining and has lost ground in chief machinery markets in some areas, such as spinning and weaving. There are no substantial American machinery producers in the market in these two areas.

The development of textile technology cannotes that productivity in the textile industry is above the average of all manufacturing industries. Unfortunately, this is not true for earnings. The technologies, however, are not to blame for the lack of performance. For instance, loom speeds have almost doubled since 1970.

Tighter international markets, imports, and the Tack of appropriate marketing are more likely the cause of unsatisfactory financial performance of the textile and apparel industry. The use of the newest technologies is not restricted to industrialized nations.

Machinery Manufacturers

Developing nations are rapidly increasing their capacities with the newest technology available. Machinery manufacturers are likely to shift their emphasis to the Far East, due to higher growth prospects. Small advantages in productivity which enable firms to be more competitive are the key to an industry which competes on a cost basis.

Labour cost differentials still are a high disadvantage for manufacturers. However, a positive aspect is that in several segments of the textile industry capital intensity is very high and labour costs become less important due to automation. New technologies in the textile industry reduce labour content, and improve productivity. In the apparel industry, the labour production component is reduced without substantial productivity increases.

The limits to labour reductions in the apparel industry are set by the sewing operation. It is feasible to almost fully automate some textile processes. One can hope that the textile industry will be more successful in becoming capital intensive than the apparel industry. The technology is available, and there are considerable developments going on in all parts of the world. Generally, apparel technology still lags substantially in automation.

The development of technologies is not at an end. Automation and robotics will further increase productivity and reduce labour. During the past decade, however, the industrialized nations, lost ground in the world markets despite increases in productivity. While capacities were only unused in developing nations up to now, they increasingly attain full capacity utilization. There is a high potential for increases in the production quantities of developing countries. Major issues on behalf of the textile competitive situation are related to the level of the technologies in the different industry segments. The question is-whether textile mills have technologies on the newest level.

THE LEVEL OF TEXTILE TECHNOLOGY

In general, the level of textile technology is considerably lower than in Europe, but both continents have superior technology levels to the developing nations. The gap, however, is fastly closing. Developing nations, especially Korea, Taiwan, and Hong Kong are spending heavily on the newest technologies at a higher rate than the Western nations. These three Far Eastern countries can no longer be counted as developing nations anymore regarding their textile and apparel industries. The weaving industry is the largest spender on new technologies. These investments are concentrated among the thin biggest corporations.

The ten largest companies account for 85% of these expenditures. It appears that the upgrading of machinery is limited to large scale operations. Apart from upgrading textile and apparel machinery, there are three traits An current development that go through all the various manufacturing processes. These are the automation of all processes, monitoring of the complete manufacturing process, and the integration of all manufacturing technologies in flexible units and fewer steps. Productivity increases and reductions in the labour content are more likely feasible through-further development of the current technologies and system-wide improvements than through expensive "futuristic" technologies like moulding garments to individuals in zero-gravity environments. Most textile machinery is imported EEC countries and Japan. Imports of textile machinery are concentrated in the fabric and yam industries as well as in some other industries.

The American market is still the most important market for all machinery producers. However, some Swiss machinery manufacturers realize the limited growth in the textile industry and are shifting towards developing countries. West Germany and Switzerland together account for two-fifths of the world exports in textile machinery. Czechoslovakia and Japan have emerged as important competitors. The concentration is rather high in the machinery industry. The Swiss based Sulzer corporation recently acquired its strongest competitor, Ruti.

Moreover, Sulzer-Ruti has close ties with Toyoda, the Japanese leader in looms as well as to British air-jet technology. These impressive market positions have been achieved with high R&D in their respective fields and traditional reliability in service. The American machinery producers basically lack any momentum in this regard. Most of the European success is also because of their service. All the big machinery producers have service facilities in the heart of the textile industry. The whole staff speaks English. If essential, spare parts can be flown in within a reasonable amount of time. The success of the American textile industry depends heavily on these machinery manufacturers. However, this increasing dependency should be a major concern of the American producers. A key to competitive textile industries is to lower down the diffusion of new technologies. This would mean to emphasize in-house, proprietary technology. It is reasonable to believe that the textile companies are unable to invest in process R&D. Furthermore, the market positions of the textile machinery producers is such that a natural dependency is inevitable.

The apparel industry, however, can take advantage from developments in the electronic industries for further applications of computer controls. American resources are enormous, if used. In general, the textile and apparel industry is largely dependent on foreign machinery which is sold all over the world. The structure and performance of the various industry segments in the textile and apparel industry vary widely. New technologies influence the structure and performance of these industries differently due to the variety of the competitive forces. Also, the structure of the industries is likely to change rapidly with the introduction of new technologies.

Past data indicates that there have been important differences between the textile and apparel industry's profitability. The textile industry's profitability has been significantly below the apparel industry's profitability. A reason for the differences in earnings might be in the higher cyclicality of the apparel industry's earnings.

Apparel purchases are more volatile to the business cycles than the primary textile industry. One might hope, therefore, higher fluctuations in apparel earnings. Earning changes in the apparel industry influence the textile and fiber producer's earnings with a time lag but somewhat smoothed out. On the other hand, there are no differences in the average productivity of labour between the two industries. Contrary to popular belief, the apparel industry is not less productive than the textile industry. This is even more interesting when the different levels of capital and labour intensities are compared.

SIGNIFICANCE OF TEXTILE INDUSTRY

The textile industries are significantly more capital intensive. As a consequence, productivity is not an argument for the observed reductions in the labour force. It is quite the escape from cost disadvantages in the U.S., thus the labour cost differential. Productivity increases come as a "free gift" with further automation and robotics. Among the textile and apparel industries, onc can observe significant differences in the levels of labour intensity, capital intensity and technological volatility. The apparel and textile industry's levels of labour intensity are different, but not as widely spread as the capital intensity levels.

Gross value of fixed assets per employee is about five times as large in the textile industry as in the apparel industry. The capital requirements are, therefore, much higher in the textile industry. Besides, this enhances the industry's entry barriers. The apparel industry's entry barriers are very low, whereas, in the textile industry, there are considerable capital related entry barriers and scale related effects. Industry concentration in the textile industry is much higher than in the apparel industry. In the textile industry, there are more large firms dominating the market, whereas the apparel industry cooprises, smaller, but so many firms.

Interestingly enough, the industry's firm size adjusted concentration, and the scale assessment levels of the two industry groups are not importantly different. When the firm size differences are ruled out, the concentrations of the four largest firms are not different for the two industries. This implies, that in both industry groups, there are a few dominant firms on the market. The respective concentration levels in both are high.

In both industries, the four largest firms account for 50% of the respective markets. The relative levels of economies of scale in the two industry groups are similar. It is not the case that in one industry chief large or small scale operations dominate. However, one can observe a wide variety of economies of scale in both industries.

Regression analysis indicates that in the apparel industry, monopoly benefits (excessive profits after reinvestments and transfers to the shareholders) are literally not responsive to changes in the labour intensity. In the textile industry, monopoly profits are negatively related to labour intensity. These results clearly indicate, that for the apparel industry, the labour intensity is not directly liable for the profitability performance.

Apparel industries that use more labour as their production factor are no better off than apparel industries that are relatively more capital intensive. On the other hand, in the textile industry, the more labour intensive production facilities are clearly worse off. As a result, other factors, for example, the increasing competition from imports, are more likely to influence the profitability of the apparel industry. Obviously, apparel industries of various degrees of labour intensity have been equally capable of protecting themselves from competition. Textile industries which are less labour intensive benefit from their technological benefit. This is reflected in the fact that they are more likely to behave in an oligopolistic manner, and thus, with higher concentration. The trend in the textile industry should, therefore, be in increased expenditures in capital to constantly upgrade the technologies. Generally, industry concentration is invariant to profitability. Firms with big market shares face the same struggles in generating enough monopoly advantages as small share firms. The tendency, however, is that the apparel industries' profitability more positively responds to industry concentration.

Thus, apparel firms with high market shares tend to be more profitable and are more successful in the battle against imports than apparel firms with small market shares. The amount of economies or diseconomies of scale has no effect on the profitability of the industry. This means that the behaviour of the industries with respect to enlarging or downsizing the operations has no effect on the profitability.

The argument that larger plants are more competitive than smaller ones does not hold in both industries with respect to profitability. It is explicit from the industry structure profiles that no textile or apparel industry is identical in the way it can and will exploit new technologies, especially automation and robotics. Not all textile and apparel industries are able to drastically improve their competitive position with new technologies. In these cases, market forces are such that imports are at levels where essentially small improvements in cost and product differentiation have only marginal impact.

This is where critical market shares have already been captured by foreign firms. The most apparent examples are in knitting and in the whole apparel sector. In the relatively capital intensive industries, like the weaving and spinning industries, automation can and will enhance the competitive position of firms which can afford it. Also, the necessary and feasible technological changes are closer at hand in the latter industries, to be beneficial within a reasonable amount of time.

Winding technologies took big steps in the past years. About full automation is or can be reached soon. The spinning process still lags in full automation. However, immense improvements are under way in the basic spinning technologies. Rotor and air-jet technologies eliminated steps in the production process. The reduction of steps in the textile production process is the most important source of cost reductions. New weaving systems drastically speed up the fabric forming process.

High segmentation of the textile machinery industry, however, hinders; efficient development of joining technologies in the sense of improving the overall textile manufacturing process. Technological developments are merely further refinement of current machinery. Completely different ways are pursued by the newly emerging textile industries, the tufting and the non-woven industries. As in every early stage of the industry life cycle, proprietary technologies are developed in-house. The profits are apparent in the impressive domestic and international competitive positions of the U.S. firms which are at least partly a result of efficient technologies. The apparel industries have yet to overcome the technological disadvantage of the needle-sewing process which needs as much labour as sewing stations. Rumors of prospective revolutionary sewing robots tend to neglect the commercial value of such gadgets. Unless these technologies are cheaper than labour, their practical value is not evident. Therefore, materials handling is where efficient technological improvements have economic value. New technologies are likely to change the economies-of scale in manufacturing of the respective industries. The weaving industry is the most clear example. Air-jet weaving technology drastically enhanced economies of scale. However, if this happens in an uncertain market environment where smaller but various production batches are postulated, flexibility might be advantageous.

The apparent tradeoff between flexibility and big economies of scale is the subject of substantial research in suitable weaving technologies in Europe and in Japan. Changes in the industry structure can be hoped, once the development of joining technologies is at the Point where spinning and winding processes, as an example, can be performed without intermediate steps. One might hope more vertical integration caused by more efficient materials flow between the spinning and the winding process. Other instances are likely to have the same effect as weaving directly from spinning frames. These ideas obviously are yet to come, but they will change the industry structure. Specialists will be less in demand and firms will vertically integrate to coordinale the workflow It is also expected that more vertical integration raises entry barriers. If this is not too late foreign competitors might find it more difficult to enter the markets.

On the other hand larger economics of scale distract from essential flexibility in the marketing and manufacturing sense. It is potent not to be undermined by quicker foreign suppliers. The degree to which automation is introduced must be carefully evaluated with respect to changing economies of scale flexibility, entry barriers and overall industry attractiveness. The mature textile and apparel industries operate with technologies which are chiefly foreign produced. Weaving and spinning machinery industries are concentrated in Europe and Japan. Diffusion of technologies can hardly be halted, such that first mover effects are crucial for the competition, but their effects are not long lasting.

LACK OF INFORMATION ON DEMAND AND SUPPLY ASPECTS

Most of the decisions a manager takes are related to demand and supply issues. But unfortunately very few are able to get it, as a result decisions taken carries risk and uncertainty. Excess inventory is one of the most common problems faced by managers which further results in long cycle-time, outdated stock, poor sale, low rates, and reduction in order visibility and finally leads to customer dissatisfaction.

Long Procurement Time

In a traditional textile industry, procurement process takes a much longer time. So, the retailers need to forecast demand and identify consumption trends at a much earlier stage. Lack of clarity about future can either result in early stock out, delay or overstock.

Supply Chain incompetency

With the urge for getting global, apparel and textiles are facing hurdles of inefficiency in carrying out various processes involved right from designing, developing samples, getting approval, manufacturing, dispatching to payment procedures. The total time taken can get extended to one year or even longer. If we calculate, production actually accounts for just ten to twenty percent of the total time. Rest of the time is taken for the information processing from one end to the other. The trajectory of development of Information Technology has intersected every application in textile industry. From enhancing performance of textile manufacturing and tighter process control, IT has inserted intelligence at every node of textile supply chain.

Step into the Global Trade

It is a fact that a company going global is opened with lot of opportunities as well as threats in terms of competition, changing trends, and other environmental changes. It necessitates managing every kind of information efficiently and at much faster speed.

INTERACTION OF INFORMATION TECHNOLOGY WITH TEXTILE SUPPLY CHAIN

Sharing of Information

Proper flow of information among supply chain member is very crucial. Such flow of information can influence the performance of overall supply chain operations. It includes data about customers and their demand, inventory status, production and promotion plan, shipment schedules, payment details, etc. Bar coding and Electronic data interchange are the two information technology tools which can facilitate information integration. Bar coding facilitates recording of detailed data by converting it to electronic form and can be easily shared among members through EDI system. EDI with its high efficiency is able to replace the traditional ways of transmission like telephone, mail and even fax.

EDI enables managers to analyse and apply it in their business decisions. It also helps in expediting order cycle that reduces investment in inventory. EDI based network enables Company to maintain quick response and closure relations with suppliers and customers, who are geographically dispersed. Manufacturers and retailers can share even new designs developed through CAD/CAM.

Supports Planning and Execution Operations

Planning and coordination are very important issues in supply chain management. The next step after sharing information is planning which includes joint design and implementation for product introduction, demand forecasting and replenishment. Supply chain members decide their roles and responsibility which is coordinated through the IT system. Various software tools like MRP, MRP-II, APSS facilitates planning and coordination between different functional areas within the organization.

Material Requirements Planning (MRP): It helps in managing manufacturing processes based on production planning and inventory control system. Proper implementation of MRP ensures availability of material for production and product for consumption at right time optimizes the level of inventory and helps in scheduling various activities. MRP system uses computer databases to store lead times and order quantity. MRP includes mainly three steps: first assessing the requirement of how many units of components is required to produce a final product; here it applies logic to implement Bill of Material (BOM) explosions. Second step includes deducting the stock in hand from gross to find out net requirement. Finally, scheduling manufacturing activities such that finished goods are available when required, assuming the lead time.

Manufacturing Resource Planning (MRPII) system is a logical extension of MRP system which covers the entire manufacturing function. This typically includes machine loading, scheduling, feedback and Software extension programmes in addition to material requirement planning. It provides the mechanism to evaluate the feasibility of a production schedule under a given set of constraints.

A textile company which has multipoint manufacturing and engaged in global business necessitates something more than MRP and MRP-II like Distribution Requirement Planning (DRP), it has ability to solve both capacity and material constraints and quickly propagates the effects of problems in both backward and forward direction throughout the supply chain. The Advance Planning and Scheduling (APSS) system includes both material focus of MRP and rapid response scheduling power of MRP-II.

Coordination of Logistics Flows

Workflow coordination can include activities such as procurement, order execution, implementing changes, design optimization, and financial exchanges which results in cost and time efficiency. The results are cost-effective, speedy and reliable supply chain operations. IT contributes towards the maximizing the value of textile supply chain through integrating supply chain operations within and outside the organization and collaborating the acts of vendors and customers based on shared forecasts.

Internet adds to IT contribution towards supply chain management through coordination, integration and even automation of critical business processes. New system of the supply chain game emerges as a result of business innovation fuelled by the Internet. Many supplying companies maintain demand data by style, size, fabric and colour to replenish inventory at retail outlet. Level of replenishing is predetermined by both parties after reviewing history of sales by product and buying behaviour of the community. New Business Models:

Data Mining and Data Warehousing

Data mining is the process of analysing data from different viewpoints and summarizing it into useful information that can be used as a basis of monitoring and control, enabling companies to focus on the most important aspects of their business. It allows users to analyse data from many different dimensions, categorize it, and summarize the relationships identified. In short it is the process of finding correlations or relationship among dozens of fields in large relational databases. Data warehousing is the repository of data and can be defined as a process of centralized data management and retrieval. Centralization of data maximizes user access and analysis.

E-commerce

E-commerce can be B2B (Business To Business) and B2C (Business To Customer). B2C commerce is the direct selling to consumers through Internet. While B2B marketplace can be defined as neutral Internet-based intermediaries that focus on specific business processes, host electronic marketplaces, and use various market-making mechanisms to mediate transactions among businesses. B2B appears to be more prospective than B2C.

E-retailing

The textile-retail giants are adding an Internet shopping-component to their offering. It has affected their distribution and warehousing infrastructure. As a result of going online, retailers have changed their supply chain strategy. High volume products with stable demand are stocked in local stores, while low-volume products are stocked centrally for online purchasing. Companies prefer a direct route to consumers by closely scrutinizing individual customer's tastes, preferences, habits, and buying patterns. Instead of waiting for consumers to visit their stores, retailers simply send them e-mails with offers. Internet has facilitated quick response system. With the use of web-enabled technology it is possible to have automatic customer replenishment system.

FUTURE OF THE TECHNICAL TEXTILES INDUSTRY

The future of technical textiles embraces a much wider economic sphere of activity than just the direct manufacturing and processing of textiles. The industry's suppliers include raw materials producers (both natural and artificial), machinery and equipment manufacturers, information and management technology providers, R&D services, testing and certification bodies, consultants, education and training organisations. Its customers and key specifiers include almost every conceivable downstream industry and field of economic activity, including the architects, engineers, designers and other advisors employed by those industries. In between lie many other interested parties, including environmental, health, safety, business and free trade regulators, patent and intellectual property agents and lawyers, investors, bankers, regional investment agencies and providers of development aid.

The task of disseminating and communicating information to all these organisations and individuals is undertaken by a growing number of specialist and general-ist publications as well as by international and local trade exhibitions, fairs, seminars and conferences. The economic importance of technical textiles worldwide therefore undoubtedly far exceeds the $60 billion.

A Changing Strategic Surroundings

If the 1980s was a period when the technical textiles industry enjoyed a rapid and increasing awareness of its existence by the outside world (as well as within the mainstream textile industry), then the 1990s was an era of more mature commercial development and consolidation as fibre producers and textile manufacturers alike concentrated on overhauling and refocusing their businesses in the wake of world recession.

The new millennium promises even fiercer international competition which will see manufacturers striving to engineer costs downwards and develop global economies of scale in production and product development. Technical textiles will become better 'value for money' than ever before and this should open the way towards further applications as existing end-uses mature.

Individual companies will become less defined by the technologies and materials they use than by the markets and applications they serve. The summarises some of the key market areas and the functions which technical textiles perform, with examples of individual products in each category. It does not pretend to be an exhaustive list which would run into many thousands of products and would constantly be changing.

3

Cleaning and Cleaning Materials in Hotel Industry

Thus we know that any establishment has to be clean, well maintained and presentable at any given moment of time. But how to ensure well maintained premises? Cleaning is the most important and primary aspect of housekeeping. It is a process of removing dirt, dust and grime by using methods such as dusting, shaking, sweeping, mopping, washing or polishing. There are certain areas you may clean daily, whereas you may clean other areas occasionally or once/twice in a year.

Since there are different types of surfaces like wall, counter tops, marble floors, ceramic tiles, wooden chairs, etc, special cleaning agents are used to clean these specific surfaces. In this chapter, as suggested, discuss these various aspects of cleaning as well as the materials and equipments used for cleaning.

MEANING AND IMPORTANCE OF CLEANING

WHAT DO YOU MEAN BY CLEANING?

Cleaning involves sweeping floors, dusting furniture and other surfaces, mopping or washing floors, polishing surfaces, articles and accessories, scrubbing tiles, sinks, toilets, disinfecting drains, rearranging cleaned areas and putting things in their specific place.

We can say that cleaning is a process of removing dust, dirt or any other undesirable materials like stains, spots, contents of an ashtray, etc. What happens if cleaning is not done on a regular basis?

- Yes, your house will become the breeding ground of insects such as cockroaches, spiders, ants, flies and mosquitoes. It will look dirty and will be most uncomfortable.
- Living in such circumstances can also lead to diseases such as asthma, bronchitis, etc. Thus, cleaning is necessary for a general presentable appearance and also to ensure good hygienic conditions. What do you understand by dust and dirt?

- 'Dust' collectively refers to the loose particles, which are very easily moved by air and settle on any surface. It is easily removed with the help of a dry cloth.
- 'Dirt' refers to dust which sticks to any surface with the help of moisture or grease. It is more difficult to remove dirt as compared to dust. Dirt has to be removed either with a detergent or any other cleaning agent. Let us now read ahead of some general methods of cleaning.

METHODS OF CLEANING

The dust and dirt can be removed by dusting, mopping etc. Based on these we can describe the cleaning methods as follows.

DUSTING

You are already familiar with the term 'dust'. But how do you remove dust? When any surface is wiped with a piece of dry cloth, (duster), it carries the loose dust with it and the process is known as dusting. This should be done with a clean soft cloth.

Fig. Dusting wiht a Cloth

SHAKING AND BEATING

What happens when you shake a cloth full of dust? Yes, the dust falls out. Similarly when you shake or beat any soft material, like a carpet/rug or a curtain, the dust falls out, making the article dust free to a large extent. This is mostly done in open air so that other things do not get dusty.

Fig. Carpet being beaten

SWEEPING

When a broom or a brush is used to carry the dust laterally along the room, the process is known as sweeping. While sweeping any vertical surface as walls, you should remember to start from the top and sweep downwards. Similarly for lateral sweeping as for floors, start from one end of the room and move to another, preferably a door, and carry the dust all along or collect in a dust pan. All the movable articles kept on the floor should be lifted, swept under, and kept back in place.

Fig. Sweeping the Floor

MOPPING

You have read that wiping with a dry cloth is dusting, similarly, wiping a surface with a damp cloth is called 'mopping'. The piece of cloth used is known as a 'mop' and is generally coarser than a duster. In this process, both the dust, as well as easily removable dirt, is also removed. Mopping is mostly done on floors. Extra attention should be paid to nooks and corners otherwise it gets tougher to remove fixed grime later on.

Fig. Mopping

WASHING

Sometimes mopping alone is not sufficient to remove dirt. Such surfaces are then scrubbed with the help of a yard (bamboo) broom along with plenty of water. Eventually the dirt loosens and is carried off by water. This process is known as 'washing'. In case of tougher stains or dirt, detergent may be added to the water.

Fig. Floor Being Washed

POLISHING

When some reagent is rubbed on a surface to bring out the shine, the process is known as polishing and the reagent applied is known as the 'polish'. Similarly, many other articles/decorative items made of brass, wood, marble etc, may be polished.

Fig. Polishing a statue

CLEANING EQUIPMENTS AND MATERIALS

Let us now learn about the equipment and other materials, which assist us in the process of cleaning. Can you name a few? Let us try to make a list of such equipments.

CLEANING EQUIPMENTS

Following are some of the equipments which you will come across during the process of cleaning:

- *Dusters*: These are mostly made of soft cotton, flannel or artificial feathers mounted on a stick. These are used to clean loose dust and are also used for wiping various surfaces. You should use separate dusters for dusting and wiping surfaces such as dining table, mirrors, kitchen slabs, etc. They should be washed and dried after use.
- *Dust pans*: These are made of either plastic or metal and have flat surfaces, rounded at the sides. After sweeping, dirt and dust is collected directly into these with the help of a broom and carried to a dustbin. Dustpans save sweeping the entire amount of dust from one room to another. Instead, dust can be collected from each room and disposed of simultaneously. Dust pans should be cleaned after use.
- *Mops*: Are mostly made of thick, loosely woven cotton cloth. These are used to wipe dust from the floors. These are dipped in clean water and squeezed before wiping the floors. You should change the water after mopping each room or when it gets dirty. You should thoroughly wash the mop and spread it for drying, after use.
- *Polishing cloth*: these are made of soft absorbent cloth such as flannel.

Dry polishing cloth helps to clean and shine the polished surfaces by rubbing them vigourously.

- *Brooms*: brooms are either soft or hard. The soft ones are used to sweep the floors, whereas the hard ones (yard broom) are used to wash the floors.
- *Brushes*: Are available in various sizes and shapes and are made of different materials. Different brushes are used for specific jobs. Brushes with nylon or plastic bristles are used for cleaning carpets or furniture, round feather brushes are used to remove cobwebs, metal brushes are used to clean wire mesh in the windows. You have special nylon brushes to clean the toilets.
- *Buckets or basins*: metal or plastic buckets/basins of suitable sizes are used to carry water, detergents and chemicals so that there are no spills.
- *Dust bins*: These are available in plastic with a lid. These should be lined with paper so that the garbage does not stick to the surface. They should be emptied and washed daily.
- *Vacuum cleaner*: It works on electricity and has a fan. This sucks in the dirt and dust from the surfaces and stores it in a disposable bag inside. This bag should be emptied regularly.

CLEANING MATERIALS

There are many materials and reagents, which help in cleaning, scrubbing and polishing surfaces. Some of these are commercial preparations for cleaning and you may be already familiar with some of them.

- *Water*: Water is the simplest cleaning reagent available to us. Some dirt may be loosened and dissolved in it. Although most of the time, some other cleaning agent is also used along with it.
- *Detergents*: Detergents are available in powder, solid (soap, soap flakes etc.) and liquid form.These are used with water to clean various surfaces. The basic ingredients in a detergent are surface active agents, known as surfactants. A detergent may have more ingredients to make it more effective, like alkaline salts, bleaches, foam boosters, germicides and perfumes. The exact nature and use of a detergent will actually vary just as to its ingredients. However, there are a few points which should be kept in mind while choosing a detergent.

  It should–

  - Be readily soluble in water
  - Be effective in all types of water and produce no scum
  - Have good wetting powers so that the solution penetrates between the article and the dirt particles
  - Have good suspending powers to suspend dislocated dirt and not allow it to settle back

- Be effective over a wide range of temperatures
- Be harmless to the article and the skin.
- Clean quickly
- Be easily rinsed away

- *Abrasives*: Some of the common abrasives are sand, finely powdered brick, saw dust, wheat bran, emery paper, fine ash, filtered chalk etc. Besides these, steel wool, nylon mesh, coconut fibres are also used to scrub dirt. Their use depends on the surface to be cleaned and the type of dirt to be removed. The extent of cleaning will depend upon the nature of the abrasive used and on the scrubbing action.
- *Acids*: Strong acids are used to clean toilets (water closet and sinks) and are available in crystals or liquid form. Milder forms of acids are also used to clean very dirty tiles. Acids should be rinsed off as soon as possible after use and should be stored away from children. Vinegar and lemon are used to clean stains on metals like brass and copper.
- *Alkalis*: Baking soda and ammonia are used as grease emulsifiers and stain removing agents.
- *Bleaches*: Stains on fabrics are removed by bleaches such as sodium hypochlorite, sodium perborate, hydrogen peroxide, sodium hydrosulphite etc.
- *Solvents*: Solvents such as methylated spirit, carbon tetrachloride, kerosene, petrol etc; are used to remove grease, wax and other stains from the surfaces. You should keep methylated spirit, kerosene, petrol, away from fire as they are inflammable. Carbon tetrachloride is harmful if inhaled.
- *Polishes*: Polishes are used on surfaces such as floors, furniture, leather and even metals. When rubbed on a surface, they provide a protective covering to the surface and produce shine. The article also gets cleaned in the process.

Ready-made polishes are expensive as compared to home-made ones. Recipes of some commonly used polishes are given below. You can easily make them at home.

But before you use polishes, you should keep the following basic principles in mind:

- Remove dust and dirt thoroughly before polishing a surface.
- Use small quantities of polish as extra polish could be harmful to the surface, besides being uneconomical.
- Rub off polishes thoroughly as surfaces could otherwise become greasy and sticky.

- Surfaces already provided with permanent or semi-permanent polishes should be polished very carefully, so as not to destroy the original shine.

Furniture Cream		
Ingredients	**Method**	**Use**
1. Bees wax 50gm	1. Shred wax and put in a pan. Cover with turpentine.	1.Use to polish light coloured furniture.
2. Turpentine 30ml	2. Heat in a double boiler till wax melts.	
	3. Cool and allow it to set.	
Furniture Polish		
Ingredients	**Method**	**Use**
1.Linseed oil 50 gm	1.Mix all ingredients together in a clean bottle and store	1.This can be applied on furniture with a pad of old cloth
2.Turpentine 30 ml		
3.Vinegar 30 ml		
4.Methylated spirit 30 ml		

Metal Polish		
Ingredients	**Method**	**Use**
1.Soap-2 tbs	1.Dissolve soap in boiling water	1. Shake well before use.
2.Ammonia-1 tbs	2.Mix with bath brick and ammonia.	2. Soak soft toweling in the mixture, let it drip dry.
3.Boiling water-2½ cup	3.Cool and store in air tight bottle.	3.Wipe and rub with this toweling.
4.Bath brick-50		
Copper Cleaner		
Ingredients	**Method**	**Use**
1.Fine sand-4 tsp	1.Mix all ingredients and keep in jar.	1.Rub well into brass or copper surfaces to remove stains.
2.Flour-2 tsp	2.Moisten a small amount of above mixture to a paste by using equal amounts of vinegar and water.	
3.Salt-1 tsp		

Note: tbs-table spoon; tsp-tea spoon

Apart from these equipments and cleaning agents, there are other materials which are used in a cleaning process, such as disinfectants, deodorants, antiseptics, etc. Can you tell where these are used?

SCHEDULE OF CLEANING

Now the important question is how to do cleaning? You must have observed the cleaning process at your own house. Do you clean your rooms completely by removing all the furniture etc, every day? No, because that would require a lot of time and labour which can not be devoted everyday. Then how to do the cleaning? For this, it is important to follow a certain schedule of cleaning. Everyday, a general cleaning of the open surfaces like floors, furniture and other such surfaces is required. Once in a while some more time is given to cleaning and you probably move heavy furniture and clean beneath it or beneath the carpets. Maybe once in six months or a year you empty the room completely and give it a complete wash, polish the floors, whitewash the walls, ceiling etc.

Thus we can basically divide cleaning into three types of schedules:

1. Daily clean
2. A weekly clean
3. A spring clean

A daily cleaning would be a general cleaning done every day; a weekly cleaning would be a more thorough cleaning done periodically, depending on the frequency of use. In a guest house, hotel, or a hospital, it may be done once a week or even earlier. Spring cleaning is usually done once a year or when particularly needed. It may be earlier in the case of a hospital.

GENERAL PROCEDURE FOR DAILY CLEANING

Let us now see how a room is cleaned daily. Can you suggest in what order the work should be carried on?

- Once you enter the room, open all windows in order to let the fresh air come in.
- Remove all unwanted articles like tea cups etc., and empty ash trays and dust bins.
- Sweep the floor.
- Dust all surfaces including furniture and fixtures.
- Brush or vacuum clean the carpet.
- Mop the whole area.
- Replace linen wherever required, like in a bedroom, make the bed, in a restaurant cover the tables, in bathrooms, check for towels, soaps etc.
- In the end, adjust windows, do a general survey to see that everything is in order and to your satisfaction.

Similarly the kitchen can also be cleaned in the same way:

- Collect all used utensils from the counters.
- Wipe and clean the gas stove, electrical appliances and the counters.
- Wash all utensils. Drain and store.
- Sweep the floor.
- Empty the dustbin, wash, wipe and line it with newspaper.
- Mop or wash the kitchen with a mild disinfectant.

GENERAL PROCEDURE FOR WEEKLY CLEANING

You now know that special cleaning is more thorough than daily cleaning.

Let us now see in what order should one work for special cleaning of a room:

- Start in the same way as in a daily clean–that is, first open all the windows for fresh air.
- Remove all unwanted articles like trays, teacups, bottles etc. Empty the ash tray and dust bins.
- Remove all dirty linen.
- Remove stains from walls, doors, windows and furniture.
- Check and clean thoroughly, all the drawers, furniture, fittings, all hangings or pictures, lights, etc.
- Wipe, dust or polish table lamps, accessories, telephone, if needed.
- Vacuum clean the carpets and other upholstery. If vacuum cleaner is not available, use a brush.
- Sweep, dust and mop the surfaces.
- Replace linen with clean linen
- Survey the room for any discrepancy and adjust windows as desired.

In the case of kitchen:

- You can empty out the shelves.
- Clean the jars and bins.
- Change the newspapers.
- Rearrange the cupboards.
- Clean the sunmica on the cupboard door panels with a wet cloth.
- Clean the tiles.

GENERAL PROCEDURE FOR SPRING CLEANING

Spring cleaning is done after long intervals, the frequency being as less as once a year. Thus, it may also be called annual cleaning. It is the most thorough cleaning of a room.

Let us now see how a spring cleaning should be done:

- Ventilate the room.
- Take off all linen, including curtains and remove them from the room.
- Remove all the movable articles including lamp shades, pictures, wall hangings etc., wipe and clean everything.

- If necessary, remove all furniture and furnishings from the room. At least remove soft furnishings like carpets. Clean the cobwebs.
- Sweep the floors.
- If any maintenance work is required, this is the right time to do it.
- Polish the furniture, decorative articles and floors.
- Clean carpets thoroughly in the sun or send for dry cleaning. Re-lay it.
- Replace everything at the predetermined place, including all furniture and fixtures.
- If desired, rearrange the heavy furniture to give a new look.
- Dust and mop.
- Adjust windows, survey the room to satisfaction.

In the case of kitchen:

- Empty out the kitchen.
- Spread the pulses, spices, etc out in the sun.
- Clean the cobwebs.
- Clean exhaust fan and light switches.
- Clean tiles with detergent.
- Spray insecticides in corners.
- Wash kitchen counters with hot soapy solution and if needed, polish them.
- Wipe stains on cupboard doors.
- Tighten any loose screws.
- Replace newspaper lining in cupboards.
- Label and arrange all boxes back in place.
- Wash the floor.

CLEANING AREAS

TYPES OF GUEST ROOMS

As far as I know, it normally starts from the Standard, Superior, Deluxe and Grand Deluxe as the very top of the categories. The names basically vary, depending on types and locations of the hotels. You may find some 5 or 6-star hotels put "Ocean, Lagoon and etc" in front of the category names, making guests realise how this room is different from others. Also, the price is higher.

SINGLE ROOM

Comfortable yet functional Single Rooms feature a 2.8-meter-high ceiling, a large writing desk and a generous-sized bed:

- *Floor Space*: 20m^2 (218sqft)
- *Bed Size*: W1,400mm×L2,030mm(55.1inches x 79.9inches)

STANDARD DOUBLE

Standard Double Rooms have an innovative yet functional layout with a diagonally positioned semi-double bed. The large-sized desk/dresser faces the window.

All Standard Double Rooms have a limited view, and look out onto Nittele Tower, the new company building of Nippon Television Network.

- *Floor Space*: 22m^2(233sqft)
- *Bed Size*: W1,400mm×L2,030mm(55.1inches x 79.9inches)

DOUBLE

Double Rooms are an excellent choice for both business and leisure travellers. All Double Rooms have a bathroom with a separately enclosed bath/ shower.

- Floor Space: 26m2(278sqft)
- Bed Size: W1,600mm×L2,030mm(62.9inches x 79.9inches)

STANDARD TWIN

The extended (2.8 meter/110.2 inch) ceiling and oversized windows promise a comfortable stay. The bathroom features a separately enclosed bath/ shower, creating a space that is highly functional.

- *Floor Space*: 26m^2(278sqft)
- *Bed Size*: W1,100mm×L2,030mm(43.3inches x 79.9inches)

BUSINESS DOUBLE

Totally equipped, comfortable and spacious room, available with a double bed or two single beds and a fully fitted bathroom with shower unit or bathtub. Following the same style as the other rooms, it has been decorated in beech wood and white colours. Bright and cosy, it is also both modern and functional.

Room services:

- Flat Screen
- Security desposit box
- National and International TV
- Air conditioning with personal control system
- Sound-proofed room
- Hair-dryer

BATHROOM CLEANING

A trip down the cleaning aisle in your local supermarket can be confusing. There are scrubbing bubbles, foaming cleansers, bathroom, tub and tile cleaners, toilet cleaners, glass cleaners, mildew cleaners--and that's just a start. You could easily spend half the day just reading labels to find out which product is the best for every single job in your bathroom. There are a few things to keep in mind, however, when choosing cleaners for your bathroom in order to make the selection process a bit easier.

Chlorine and ammonia don't mix. Or at least, they shouldn't. When chlorine and ammonia get together, a chemical reaction releases harmful and potentially toxic fumes. This is important to know because ammonia and chlorine bleach are the two most common bases for many power-scrubbing bathroom cleaners. When you mix and match bathroom cleaners, you increase the likelihood of accidentally combining ammonia and chlorine. If possible, stick to either ammonia based cleaners or bleach based cleaners only and never combine the two. One way to avoid the potential for mixing the two chemicals is to choose a multi-surface bathroom cleaner that you can use on multiple surfaces. If you can find one product that removes hard water scale, eliminates soap scum, and can be used on porcelain, Formica, glass and mirrors, tiles, and many other surfaces, you'll eliminate a potential hazard and save money at the same time.

You can reduce the work of cleaning the tiled walls of your shower by applying a coat of car wax to them after they have been cleaned. Make sure you only wax the walls and not the floor of the tub, as the wax can make things awfully slippery. Glass and fibreglass are both porous, which makes them prone to staining from hard water. The car wax seals the pores, however, making them less prone to staining.

General Duties:

- Rinse out the sink after each use.
- Hang up towels and washcloths.
- Wash out the bathtub after each use.
- Flush the toilet after each use.
- Remove excess hair from the sink or tub.
- Remove dirty clothes.

Supplies needed:

- All-purpose cleaner
- Cloths or sponges
- Toilet brush
- Glass cleaner (for mirror)
- Mop and bucket
- Rubber gloves

All-purpose cleaner:

- Mix 2 tablespoons ammonia, 2 tablespoons liquid dish detergent, and 1 quart of water.

To clean the sink, faucets, tub or shower:

- Wash with a soapy cloth or sponge.
- Use baking soda and a nonscratch pad for stubborn marks.
- Rinse with clear water.

To clean the toilet:

- Wash the tank, outside of bowl, seat, and lid with a soapy cloth.
- Scrub the inside of the toilet bowl with a long-handled brush.
- Any all-purpose cleaner will work if the toilet bowl is cleaned weekly.

USE A BATHROOM STEAM CLEANER FOR A HEALTHY, SPARKLING ROOM

The average person can ignore a little dust, but when it comes to a grimy, dirty bathroom that is a completely different story and where a bathroom steam cleaner comes in handy. Nobody likes to use a bathroom that looks unkempt and dirty. At the very minimum, you should clean your bathroom at least once a week and even more often if you have a large family or the bathroom gets a lot of use. There are a few tips to follow daily that make your bathroom easier to clean weekly.

After a shower wipe the shower stall, walls or tub with a damp, soapy sponge and rinse. Rinse your sink after each use and always clean up any spills immediately. A steam cleaner comes in many different sizes, shapes and models and uses steam vapour to deep clean, deodorize and sanitize in one easy step. There are vapour steam cleaners and dry vapour steam cleaners that produce low moisture and high temperature vapour.

A vapour steam cleaner does not use chemicals for cleaning so it is a healthy way to make your bathroom shine. For allergy sufferers, a bathroom steam cleaner penetrates the object that you clean and kills any molds, mildew, fungi, viruses or dust mites instantly, which improves the air in your bathroom immediately. The wonderful thing about this type of cleaner is it is useful in every room of your home, not just the bathroom.

It is also an environmentally safe and fast way to clean your home. A steam cleaner uses very little water, as the vapour produced is low moisture, so it leaves almost no damp residue on the cleaned surface. This is especially handy if your bathroom has carpeting because, after steam cleaning, your carpets will dry completely in approximately ten to fifteen minutes.

Once it is dry, vacuum the carpet to remove any particles of dirt loosened. Using the steam cleaner on hard surfaces dislodges the dirt and grime, which you just wipe up with a cloth. The bathroom steam cleaner works great around the bathroom window, on grout and kills the mold spores.

In the bathroom, it also cleans many other things such as:

- Faucets
- Floors
- Shower stall doors
- Bath surrounds
- Inside and outside of the toilet
- Tiles

- Shower curtains
- Walls and ceiling

By cleaning your entire bathroom with a steam cleaner it not only smells clean and fresh but removes germs, viruses, bacteria, mold and odours, making it very sanitary. Using a bathroom steam cleaner will make your bathroom sparkle, shine and look clean and fresh in no time.

ORGANIZE YOUR BATHROOM CLEANING SUPPLIES

The bathroom cleaning supplies needed to clean a bathroom are as numerous as the different surfaces and varieties of dirt and scum, which form in the area. There is some or all of the following surfaces in many bathrooms such as porcelain, glass, mirror, chrome, tile, fibreglass and ceramic. Some of the soils that need cleaning are soap scum, dirt, rust, hard water deposits, mildew, stains and mold. Because this can become overwhelming the best way to tackle cleaning a bathroom is by being organized and having the necessary bathroom cleaning supplies.

When your supplies are organized and together in one place, it makes cleaning the bathroom so much quicker and easier. Go through all your cupboards and drawers and gather the bathroom cleaning supplies you use specifically for your bathroom cleaning together. Throw away any products you have had for many years and never use, as they probably lost a lot of their effectiveness. Toss any that are almost finished or you do not use anymore, along with old, worn out, sponges and rags. Find a large bucket or shelf and fill it with all of your cleaning supplies.

Keep it well stocked with items you need to make your bathroom clean, fresh and sparkling. Some of the items that you may use or need on a regular basis, for cleaning your bathroom, are window cleaner, sponges and rags, toilet, sink and bathtub cleaner, paper towels, duster, floor cleaner, tile cleaner and a good all-purpose cleanser. Finding or making a cleaning product, that does more than one job, cuts down on the amount of different cleaning items needed and this saves you time in the end.

You do not need to load your shopping cart with cleaning supplies for your bathroom, as there are many great, simple, inexpensive recipes for making your own effective cleansers at home. Many are more environmentally friendly and do not contain irritating fumes and harsh chemicals. Stock the cleaning supply bucket or cupboard with homemade cleaning solutions and sprays, to get rid of household dirt and grime.

Here are a couple of great cleaning products to add to your supplies:

- In a spray bottle, mix a cup of white vinegar and one cup of water. Use this spray to clean the outside of the toilet, floors, countertops and shower stalls. For tough bathroom dirt, mix a cup of white vinegar and a cup of hot water. Spray this heated mixture on, leave it for ten minutes, so it softens stubborn hard water deposits, and soap scum, scrub and then rinse. Undiluted white vinegar also works great for cleaning toilets, showerheads and dissolving hard water deposits.

- Make a great, inexpensive glass and mirror cleaner by mixing one cup of rubbing alcohol, a cup of water and one tablespoon of white vinegar in a spray bottle. Simply spray on and wipe off with a lint-free cloth or paper towels. You can also replace the white vinegar with a tablespoon of non-suds ammonia instead, for a tough, all-purpose cleaner.
- To keep bathroom drains clear, pour a cup of baking soda down the drain, followed by a cup of white vinegar and leave this for an hour. Flush with warm water. This also works for kitchen sinks, bathtub and shower drains.

Cleaning is not most people's idea of fun but by keeping your bathroom cleaning supplies organized you will finish the job faster and have more time to relax and enjoy your day.

BATHROOM TILE CLEANER THAT GIVES YOU SPARKLING RESULTS

Knowing what type of bathroom tile cleaner to use on the different types of tile and different types of tile problems is often confusing for people.

Cleaning a bathroom, including the bathroom tiles, is a household task that most people dislike. Because of this, they do not clean the bathroom often enough and the dirt and grime build up, which makes the job of cleaning it even more difficult when the time finally arrives. Most of us can tolerate a little dust occasionally, but when it comes to the bathroom, everyone dislikes going into that room and feeling like it is not clean and fresh.

It can be embarrassing to have company drop in and instead of enjoying yourself, you worry about someone wanting to use the washroom and noticing your stained, marked tile floor. Its time to get organized, learn some cleaning tips and also what bathroom tile cleaner to use so you have a bathroom that shines and looks fabulous. Start by gathering all your bathroom cleaning supplies.

Put them in a large bucket or on a shelf for easy access whenever you clean. Make sure you store them in a place that is child and pet safe. Remove the rugs and garbage pails, so you do not have to work around anything. Sweep your tile floor before you start to clean the bathroom so if water drips on it you do not end up with a muddy floor. Clean everything in your bathroom except the tile floor and the shower and shower floor or bathtub walls and bathtub. Always work in a well-ventilated area and wear rubber gloves when cleaning. Here are a few great bathroom tile cleaner products that are inexpensive to make, easy to use and work great for bathtub, shower and sinks.

- Combine one cup of clean ammonia, a quarter cup of baking soda, a half-cup of white vinegar and a gallon of warm water. Apply the cleaning solution, let it sit for a few minutes to dissolve the dirt and grit and then wipe off and rinse.
- In a gallon of warm water mix, a half cut of vinegar, one cup of ammonia and a cup of baking soda. Use a mop to clean the tile floor with this mixture. In the bathtub, sponge on or spray, leave it sit for a few minutes and rinse off.

If you have a grout problem, use an old toothbrush or grout brush and scrub with a solution of five parts water to one part liquid chlorine bleach. Let it sit for a few minutes and rinse with warm water. Repeat if necessary. Using a great bathroom tile cleaner and cleaning a little more often, will have your bathroom clean and sparkling all the time.

CERAMIC TILE CLEANING LETS THE TILES BEAUTY AND LUSTER SHINE THROUGH

With a little ceramic tile cleaning your ceramic floors and tiles will look as lovely today as they did the day you had them installed. Ceramic tiles come in two basic types, which are unglazed and glazed. Along with cleaning the ceramic tiles you must give some attention and care to the grout between the tiles. Grout gives your floor or walls a finished look, as it fills the spaces in between each tile. Grout comes in several colours and types but they all require cleaning and maintenance to remain in good shape.

They often use unglazed ceramic tiles on floors and countertops. Unglazed tiles require more attention than a glazed ceramic tile does. They traditionally use glazed tiles on floors, walls and countertops and these tiles have a tough, almost glasslike surface. Ceramic tiles are available in many finishes such as matte, glossy or textured. Ceramic Tile Cleaning For Unglazed Tiles Always sweep or vacuum your unglazed tile floor regularly to remove any dirt or gritty particles. Wash the floor with a mild soap-free detergent and warm water. You can also damp-mop unglazed tiles with a gentle all-purpose cleaner and dry with a soft cloth to avoid any streaking. Use the same solution on a sponge, for cleaning unglazed wall tiles.

Never use strong soaps, abrasives or acids on unglazed ceramic tiles. You should seal unglazed or matte ceramic tiles with a commercial sealer, as they are very porous and stain. On any ceramic floor, wipe up spills immediately to avoid staining. Ceramic Tile Cleaning for Glazed Tiles Sweep or vacuum your floor before cleaning. Damp mop the floor, using warm water and a soap-free mild detergent. Rinse with clean, warm water. For walls and shower walls, a mixture of equal parts white vinegar and warm water often removes soap scum and dirt. Use chlorine bleach and water to clean dingy, dirty grout. Apply this solution, scrub with a grout brush or toothbrush and rinse. Use this only on light, non-coloured grout. You can also use a mixture of fifty-per cent hydrogen peroxide and fifty-per cent warm water in a spray bottle to clean grout. On stubborn stains, spray on shaving cream, leave for a few minutes and rinse off. Never use anything abrasive, such as scouring powders, on your ceramic tile as they could scratch the finish. By regularly ceramic tile cleaning and wiping up spills immediately you will keep your floors, countertops and walls looking wonderful.

MAKE YOUR BATHROOM MILDEW FADE AWAY

Walking into the bathroom and smelling a sour or musty odour is very upsetting and unpleasant and more than likely bathroom mildew and mold are the culprits. Mold and mildew are fungus that grow anywhere in a home but they especially love bathrooms. The reason for this is that it meets their environmental needs, which are poor or low air circulation and ventilation, low-lighting, high humidity, dampness, and warm temperatures. Mildew and mold grow on almost any surface such as tiles, grout, paint, wood, glass, material and porcelain.

At one time or another, most small bathrooms end up with mold and mildew. Bathroom mildew finds a damp, dark crack or corner on the wall or in the shower, slowly starts growing and before you know it, you have a big problem on your hands. It continues growing, attaching itself to the grout between the tiles, the silicone sealer and even on the ceiling. Without getting the mold and mildew under control, your beautiful bathroom could end up looking and smelling terrible.

Controlling Mold and Mildew

Controlling the humidity in a bathroom, caused by running water and steam, is extremely important. Turn on your bathroom fan to pull steam and moisture from the air and if you have a window, open it a bit. This helps to keep the air circulating. If you do not have a bathroom vent or fan, it is a great investment but be sure it vents directly outside. Mildew loves dark areas so turn on your lights before a shower or bath and leave them on for an extra fifteen or twenty minutes after.

Wipe the bathtub or shower area with either a squeegee or dry cloth after bathing and always hang up damp towels and clothing. Use either bleach or vinegar regularly, to stop future growth of mold and mildew. When the time comes to repaint, use a mildew proof paint, designed specifically for bathrooms, kitchens and other high-moisture areas

Removing Bathroom Mildew

Always wear rubber gloves and work in a well-ventilated area when cleaning. A mixture of five parts water to one part chlorine bleach is great for killing bacteria as it cleans. Use this on all the bathroom surfaces. In the shower or bathtub areas, use a stiff brush to get into the grout, tiles and hard to reach corners. If you have a corner that is difficult to reach even with a brush, soak a cotton ball in bleach. Place the cotton ball against the mildew, for twenty minutes, remove and then rinse.

Wash your shower curtain in the washing machine but add a cup of bleach to kill any mildew or mold. Once your shower curtains are clean, soak them in salt water to prevent future mold and mildew growth. Keep a spray bottle of vinegar in your bathroom and after each shower or bath, spray the walls and tiles around the bath area. Clean your bathroom often with bleach, vinegar or commercial bathroom mildew cleaner to keep your room sparkling fresh, mildew and mold free.

CLEANING TOILETS, FLUSH THE DIRT AWAY

Cleaning toilets is not a job anyone looks forward to but it is very important not only for the appearance of your bathroom but also your health. As hard as it is to believe, there really is a correct and incorrect way to get your toilet sparkling.

People spend far too much time worrying about the inside of the toilet bowl and too little time on the outside of the toilet. The outside of your toilet is where you need to spend the time cleaning and disinfecting because you sit on the seat, lift the lid and push the handle. The parts of the toilet you touch are more important than the toilet bowl. One thing you should never do is flush the toilet with the lid up, as it sends bacteria everywhere.

When tackling the job of cleaning toilets gather all your supplies together in one place. This makes it far easier and quicker to get started. Some of the supplies you need are rubber gloves, a stiff plastic bristle brush and holder, paper towels, cleansing powder, bleach spray and all purpose cleaner. You can make your own bleach spray by mixing 3/4 cup of bleach and a gallon of water. Fill a spray bottle with this solution. Make a good all-purpose cleanser by mixing a cup of rubbing alcohol and a tablespoon of ammonia in a spray bottle and then fill up the rest with water.

Wearing your rubber gloves, always start cleaning toilets by pouring or sprinkling cleanser into the toilet bowl, but not in the water. After leaving it sit for a few minutes, use your toilet brush and scrub, starting under the rim and clean to the bottom of the bowl. Flush your toilet with the lid down, when you finish scrubbing. Spray the outside of your toilet with an all-purpose disinfectant spray.

Using paper towels, start wiping the toilet down, beginning at the top of the tank. Pay special attention to the handle area and the toilet base. Now, spray the toilet seat outside and inside, the hinges and then the rim of the toilet. Using the paper towels, wipe this area clean, from the lid to the toilet rim. Always start at the top and work your way down. Nobody likes to use a bathroom that looks like it needs a good cleaning. Clean your bathroom weekly or more often if used a lot. Pay special attention when cleaning toilets and your bathrooms will sparkle and shine.

CLEANING BATHTUBS AND HOMEMADE BATHTUB CLEANERS

Cleaning bathtubs more frequently and using a great bathtub cleaner makes the job of cleaning your bathroom far easier and faster in the end. Leaving your bathtub and only cleaning in every few weeks is more time consuming and difficult than cleaning it once or twice weekly.

Some of the supplies you need to clean your bathtub and bathroom are rubber gloves, mildew remover, bathroom disinfectant cleanser, old toothbrush or a grout brush, soft-bristled brush, and rags or a sponge. You should clean the walls or surround above your bathtub before you start cleaning bathtubs Spray an all-purpose cleaning solution or your homemade solution on the walls and then leave it for five minutes, until it sinks in and loosens the dirt and grime.

While wearing rubber gloves, gently scrub any built-up dirt or soap scum and then rinse. If the grout needs cleaning, spray with a mildew remover or make your own. Scrub with either a grout brush or old toothbrush, to remove the grime and mildew. Rinse the walls and wipe dry.

When cleaning bathtubs, spray the tub with a bathtub cleaner, until you coat the tub and then let it sit for five minutes. This gives the cleanser a chance to sink in and loosen any dirt or grime. Using a soft-bristled brush, gently scrub out your tub and then rinse well. Use an old toothbrush to get at hard to reach areas, such as the corners, around the taps or the grout around the tub. Rinse well when you finish cleaning your bathtub.

Here are a few suggestions and tips for cleaning the bathtub:

- After taking a bath, on a net bath puff, pour some inexpensive shampoo, wipe around the sides and bottom of the bathtub and rinse. This is an easy, quick way to keep your bathtub shining.
- Sprinkle borax in a wet bathtub and then scrub and rinse.
- Mix together 1 teaspoon of vinegar or lemon juice and a quart of warm water. Put this in a spray bottle and use to clean your tub, sink and other surfaces.
- Make a paste of vinegar and baking soda to scrub the tub and shower. This removes mildew and soap scum easily but be sure to let it sit for a few minutes to soak in. Rinse with warm water.
- Wearing rubber gloves, apply straight peroxide to light coloured grout and leave this sit for five minutes. Use a grout brush or old toothbrush to scrub the grout and then rinse with warm water.

Cleaning bathtubs more often and using the correct bathtub cleaner will keep your bathrooms will sparkling and shining.

THE EIGHT TOP BATHROOM CLEANING TIPS

Nobody likes to clean the bathroom but by following a few bathroom cleaning tips and suggestions, this job can certainly be faster and easier. A grimy bathroom is something most people cannot tolerate. Cleaning it weekly or more frequently, if you have a large family, does not have to become an all day job. Fortunately, most of the materials and surfaces in bathrooms are not too difficult to clean. The real problems start when you neglect the bathroom for a long time and the scum and dirt have a chance to build up. That is why it is extremely important to clean your bathroom weekly at the very minimum.

Here are a few great cleaning tips that will have your bathroom looking clean and fresh in no time:

- In a large spray bottle, use a mixture of 1/4 cup of rubbing alcohol, 1/3 cup of white vinegar and dilute this with water. Spray this on your bathroom mirrors and windows, wiping with a paper towel or even newspaper, and they will shine.
- Here are a couple of great bathroom cleaning tips to remove built up soap scum and grime from your glass shower doors, spray oven cleaner on the inside of the door. Leave it from half to three quarters of an hour. Scrub any built up dirt areas and then rinse with warm water. This also works fantastic on mildew. Dry the doors using a squeegee.
- A lemon cut in half and then rubbed on the glass shower doors, cuts through the dirt scum.
- To clean your tiles, mix three cups of water and three-quarters of a cup of bleach in a spray bottle. Spray, leave for a few minutes and wipe off.
- Baking soda works great for cleaning fibreglass bathtubs and shower stalls. Just sprinkle onto a damp surface, scrub and rinse. This mixture is safe to use on your bathroom facets also. For stubborn stains, a Mr. Clean magic eraser works.
- Do not be afraid to throw your vinyl shower curtain in the washing machine on delicate or gentle, with a small amount of bleach and cool water. It controls mildew and your shower curtain lasts longer.
- Wash any throw rugs in your bathroom weekly, when you clean.
- Pour some bleach into your toilet bowl, let it sit for a while and scrub with your toilet brush. Flush and it is clean.

ELIMINATE BATHROOM ODOUR COMPLETELY

It is very upsetting to walk into your bathroom and smell an offensive bathroom odour. Some of the causes of the odour are clogged drains, mold and mildew, a kitty litter box or just a plain dirty bathroom. Here are some suggestions and tips that will have your bathroom sparkling and smelling great-quickly and inexpensively in no time.

Clogged Drains

It is no wonder that bathroom drains occasionally clog up with all the hair, dirt, soap and other gunk that goes down them. A good preventative maintenance tip is every three or four weeks, pour a cup of baking soda down your drains and then a cup of white vinegar. After an hour, turn your tap on and run some warm water through the drain.

If your drain is removing the water slowly from your sink, plug the overflow openings and then use your plunger to unclog the drain. If the water is still very slow draining-there are strong chemical drain openers on the market. Be sure to protect your eyes and skin and only use this in a well-ventilated area. Follow the manufactures directions. If your drain is clear but emits an odour, pour either bleach or vinegar down the drain.

Cat Litter Box

To eliminate the odour from a cat litter box, rinse the litter box and pour some white vinegar onto the bottom. Leave for approximately fifteen minutes, rinse with cold water and then dry with paper towel. The vinegar neutralizes the smell of ammonia, from the litter. Add baking soda to the kitty litter to control the odour. There are also many litter boxes, on the market, that eliminate cat litter odour.

Mold, Mildew and Dirty Bathrooms

Instead of walking into bathroom odour when you enter the bathroom, eliminate problems such as mold, mildew and dirt. You will have it smelling fresh, and clean quickly and easily. A great all-purpose cleaner to use on most bathroom surfaces is a gallon of warm water with a half cup of bleach mixed into it. This disinfects as it cleans.

NEVER mix bleach and ammonia together. Use this on your bathroom ceiling, walls, floor, tiles, toilet, sink, and bathtub. For mold and mildew, spray one part bleach and five parts water onto the grout and repeat until the mold and mildew disappear.

zPutting an air freshener in your bathroom may mask the bathroom odour but you need to find the cause and eliminate it so your bathroom smells as great as it looks.

SIX GREAT SHOWER CLEANING TIPS

Shower cleaning is definitely necessary to keep this important area of your bathroom clean and sparkling. Dirty or moldy looking walls in the shower make the whole bathroom feel unsanitary.

Here are some shower cleaning tips that will keep your shower and walls, sparkling and fresh looking. Always wear rubber gloves and keep the bathroom well ventilated when you clean.

Never mix chlorine bleach and ammonia, as these two products together produce very toxic fumes:

- Always wipe or squeegee your tiles or walls down after you take a shower or bath. Mold and mildew love damp, warm places so the dryer you keep your tiles, the less chance that mold and mildew will try to take over.
- Open the bathroom window slightly to circulate the air while taking a shower and turn on the bathroom fan to remove moisture. Keep a light on while taking your shower and turn it off about twenty minutes after you finish.
- Make a mixture of one part chlorine bleach to five parts warm water and wash the shower ceiling, tiles and floor with this.
- Another very good shower cleaning solution is a half-cup of ammonia and one gallon of water. Put this mixture into a spray bottle to use on your shower doors. This works great for removing soap scum. Spray the solution on your glass shower doors, scrub with a stiff-bristled brush and the rinse well with warm water.
- Shower door tracks get very dirty and need special attention occasionally. Start by removing any old hair and dirt from the shower tracks, using an old toothbrush or a small, stiff brush. Vacuum this up with a long-nosed vacuum attachment.
- Pour undiluted white vinegar into the track and leave for approximately fifteen minutes. Using a small stiff brush or old toothbrush, scrub the track and be sure to get into those hard to reach places. Rinse with warm water and repeat, if necessary.
- Use liquid car wax on clean fibreglass shower walls and glass doors to prevent alkaline deposits. It also makes cleaning the shower easier and prevents water spots. Reapply this mixture every six months.
- Wash your shower curtain in the washing machine but add one cup of bleach and one cup of vinegar to the wash water. Throw in a couple of white towels, laundry detergent and wash in warm water. Hang the shower curtain back on the shower rod to dry. You can also substitute a cup of baking soda for the vinegar and bleach.

FIVE TIPS FOR CLEANING BATHROOM TILE

The bathroom is one room of the house that should be kept as clean as possible and using a few easy tips can help with cleaning bathroom tile. Bathroom tile that is not clean is dull and stained looking and is a breeding ground for bacteria, mold, mildew and even fungus.

Keeping your bathroom tile clean is easier if it is done on a regular basis, ideally daily, instead of waiting until there is a noticeable build-up before cleaning.

For daily cleaning of bathroom tile consider these five easy tips:

1. Cleaning daily is the key to making your tile sparkle. If your shower area is tiled simply mix 3/4 of a cup of bleach and 3 cups of water and keep in a spray bottle in the shower. After every shower simply spray down the walls of the shower with the mixture and wipe off with a sponge. The bleach will eliminate soap scum and will also prevent any mildew from growing on the tile or the grout.
2. For heavier build up or stains mix 1 cup of salt, 1 cup of baking soda and one cup of vinegar together to form a thick paste. Apply to the stained or soiled area with a soft cloth and allow to stand for three to five minutes. Rub in a circular motion and rinse off with clean water. Be sure to use a non-abrasive sponge or cloth to avoid scratching the tile's surface.
3. Keep your bathroom as dry as possible. While this is not a specific cleaning technique it will help prevent the build-up of mildew and mold. Keep lights on and run the vent fan while showering. Even using a hair dryer in the bathroom will help decrease the humidity in the air. Never leave the bathroom door closed after a shower as this will keep the humidity high and provide ideal conditions for mold and mildew growth.
4. To eliminate the need for constant cleaning of white grout between tiles use a combination of one or two tablespoons of bleach to one cup of water. Spray and let stand, then wipe dry with a cloth. Once the grout is completely dry apply a grout sealer to the area between tiles. This will keep mildew from developing and keep grout looking bright and wipe. Do not use this solution on coloured grout; rather use a non-bleaching cleaner and water.
5. Ammonia and water can be mixed, in a one part to three part ratio and sprayed on bathroom tile and shower doors to remove soap scum and leave a shiny surface.

Cleaning bathroom tile doesn't have to be difficult if you use a daily spray regime and keep grout and tile mold and mildew free.

CLOROX BATHROOM CLEANER TO THE RESCUE!

Clorox bathroom cleaner is a versatile cleaning product that can be used to keep you bathroom looking clean, fresh and shining. In addition it will prevent the growth of mold, mildew, bacteria and fungus that can cause health problems if left unchecked. Clorox bathroom cleaner is different from many products in that it does not contain bleach so is safe to use on most the surfaces in your bathroom. It is always recommend to test a small area of the surface you will be cleaning for the first time. Generally test under the lip of the sink to test the countertop, or inside the cupboard to test the cabinet. Clorox bathroom is not recommended for use on varnished wood, natural marble or brass as it can cause discolorations.

This cleaner works on:

- Soap scum
- Dirt
- Hard water residue
- Mold and mildew
- Bacteria

Clorox cleaner, just like all cleaners, works best on surfaces that are mildly to moderately soiled or stained. For very heavily stained or dirty bathrooms there may need to be more than one application of the cleaning product. Generally to use Clorox Bathroom cleaner simply spray the cleaner using the foam setting on the surface. Keep the nozzle of the bottle between six to eight inches from the surface and try to avoid missing any patches. Allow the foam to sit on the surface for a few minutes and then wipe off with a wet sponge or cloth.

One of the many benefits of Clorox bathroom cleaner is that the foam is thick and sticks even to vertical surfaces in the bathroom. This makes cleaning shower walls, doors and even the backsplash on sinks simple and easy. Using the fine sprayer adjustment is a great way to shoot the cleaner just where you need it, even around the base of toilets or behind taps and drains.

Keeping even the unseen parts of the bathroom mold and mildew free is important to prevent it from spreading. A new variety of Clorox bathroom cleaner now comes with Teflon that is designed to minimize the amount of substances that stick to the surfaces of the bathroom. With use the Teflon does prevent mold and mildew growth as well as minimizing soap residue and hard water grime that adheres to the shower surfaces. When using Clorox bathroom cleaner with Teflon be sure to have a fan circulating air and prevent using it in closed or confined spaces without appropriate ventilation.

CHOOSING THE RIGHT CLOROX TOILET BOWL CLEANER

There are several different types of Clorox toilet bowl cleaners come in many different types to allow you to choose the one that is right for your individual cleaning needs. Cleaning the toilet is usually one of the least favourite of all the bathroom cleaning chores, so using the right product is important to make the job easy and effective the first time. To understand the features of the Clorox toilet bowl cleaner family the following can be used as a guideline: Clorox toilet bowl cleaner with bleach-for that deep down clean Clorox toilet bowel cleaner with bleach provides both cleaning and disinfecting action.

Simply flush the toilet and apply about six ounces of the product under the rim of the toilet. Using a toilet brush scrub the area above and below the water line and let stand for 10 minutes. This allows time for the bleach to disinfect the toilet and kill the bacteria. Flush to rinse. Clorox toilet bowl cleaner with bleach and rain scent-provides all the benefits of the original Clorox cleaner without the harsh scent. Perfect for using to keep your bathroom smelling clean without chemical fumes or smells. Clorox toilet bowl cleaner with Teflon-provides all the cleaning power of Clorox with the added non-stick benefits of Teflon. Use this product with correct ventilation, as it is a bit stronger smelling than the original Clorox.

Clorox dual action toilet bowl cleaner-two powerful cleaners are mixed upon use and act to clean and disinfect the toilet. Use the same as the original cleaner but be sure that both openings are directed under the rim. With all Clorox products it is important to use some general safety measures when using the cleaners. Never mix Clorox cleaners with any other chemicals or cleaners including bleach. The mixture of various cleaners can cause the formation of potentially toxic gases that can be deadly. Avoid using any cleaners in closed or confined spaces, always use a circulating fan or open a window when applying cleaners. Clorox also makes other toilet cleaners such as automatic cleaners that sit in the back of the toilet tank to continually clean the toilet between scrubbings, and Clorox toilet wand systems that combine both the cleaner and the scrubbing brush, all in one easy to use and disposable package. Try experimenting with the various Clorox toilet bowl cleaner products and determine which ones best meet your needs.

TOILET CLEANERS FOR A PRETTY POTTY

The smell of a dirty toilet could be your greatest motivator for cleaning it as quickly as possible.

Here are some tips before we get started:

- Always wear rubber gloves when cleaning your toilet.
- Never combine ammonia and chlorine bleach as it produces a toxic gas.
- Always clean in a well-ventilated area.

HOMEMADE TOILET BOWL CLEANERS

- To remove hard water stains from your toilet bowl, start by turning off the water shut-off valve to the tank so when you flush the toilet you remove most of the water from the bowl.
- Fill a spray bottle with white vinegar and use this to spray the entire toilet bowl.
- Sprinkle borax in the toilet over the vinegar and then let this mixture sit for about half an hour.
- Using a stiff bristled toilet brush, scrub the inside of the toilet, adding more vinegar if necessary. Turn the water back on and flush the toilet.
- Dissolve a couple of denture-cleaning tablets in the toilet and leave this overnight to help dissolve hard watermarks and stains.

- To remove rust stains from your toilet bowl, use two tablespoons of tang and sprinkle on the rust stains. Leave on for approximately an hour and then scrub with your toilet brush. Flush and repeat if necessary.
- Two great toilet cleaners are rubbing alcohol and borax. Mix one gallon of hot water and a half cup of borax and use this solution to clean the outside of your toilet.
- It not only cleans but also deodorizes and disinfects. Put rubbing alcohol on a sponge and clean your toilet handle, lid and seat to disinfect. Allow this to dry and do not rinse off.
- Pour one-half cup of baking soda onto the sides of a wet toilet bowl. Let this sit for half and hour, brush and flush.
- Pour a half cup of bleach into the toilet bowl, leave overnight and scrub in the morning. Flush and you have a clean toilet.
- Pour one-half can of coca cola into your toilet bowl and leave it for a few hours. Relax and drink the other half can. Scrub with a toilet brush and then flush.

CLEANING MOLD TIPS

Cleaning mold needs to be done by a professional as mold is spread through tiny spores which once opened, float throughout the home. Once they get into the carpeting or even the air duct system-the spores are blown throughout the inside of the home and can cause harsh allergy symptoms. Allergists are finding more and more people who live in a damp climate are suffering from aches, pains, fatigue, and constant congestion due to the non stop growth of mold. Today mold is considered a dirty and toxic word. No one wants to have any kind of mold in their home. Usually mold grows in dampness and common place is in the bathroom and shower. There are cleaning products on the market which claim to kill the mold by spraying it on the infected areas and let sit for a few minutes before cleaning the mold away. So-how do we actually know how to clean mold? To date, many people believe that by using straight bleach, bleach mixed with water, or a bleaching product supposedly destroys the growth of mold. Unless you know of a professional mold cleaner, always consult your town's hazmat board and ask them to come inside the home to evaluate any possible mold growth.

MAKE LIME SCALE AND WATER MARKS DISAPPEAR IN YOUR BATHROOM

Anyone that lives in an area with hard water knows how difficult it is to remove lime scale buildup and watermarks in the bathroom. Even in a clean bathroom the unsightly mineral buildup and stains in the sink, toilet, tub, shower and on the tiles and fixtures caused by hard water makes it look unkempt and dirty. If you have hard water, you know that lime scale deposits are so hard that you need a chisel to remove it. One way to eliminate this problem is by installing a water softener but that does not get rid of the lime scale buildup and watermarks that are already there.

Here are some ways to remove the lime scale from your bathroom:

- Showers and Tiles–Use a heavy-duty gel cleaner made specifically for removing lime scale buildup.
- Avoid abrasive cleaners as they can scratch tile, porcelain, and other finishes.
- Be sure to leave the window or bathroom door open for ventilation and wear rubber gloves.
- Use a damp cloth or sponge and apply the cleaner to the tiles, shower door and other surfaces. Leave it on for the length of time the manufacturer suggests and then rinse thoroughly.
- Soak the showerhead in white vinegar for a few hours to remove lime buildup.
- Sinks–You can use the same heavy-duty gel cleaner on your bathroom sink and faucet.
- Use an old, soft toothbrush to apply the cleaner behind and around the faucet. Use warm water to rinse it thoroughly.
- Toilet–There are several liquid and gel cleaners for removing lime scale made specifically for toilets.
- Put this around under the rim and on the scale, leave it for the suggested length of time, scrub with a toilet brush, and then flush a few times.
- Another solution is to use a mixture of half borax and half vinegar. Drain the toilet bowl, apply the mixture to the lime scale and leave it on for a few hours before removing.
- Another solution for removing lime scale buildup and water marks is taking an empty spray bottle and filling it with a half cup of white distilled vinegar and two cups of water.
- Spray this on all the problem areas and wait for an hour or so but be sure not to let it dry. Spray again if necessary.
- Rinse all the problem areas with warm water and repeat if necessary.
- If the lime scale buildup is very heavy, soak some extra-strength paper towels in white distilled vinegar or all-purpose bathroom cleaner.
- Drain the toilet and line the bowl with these soaked paper towels.
- Do the same with the sink and shower if necessary.
- Leave them for several hours but add more vinegar or cleaner if the paper towel starts to dry out. Remove the paper towels, discard, and scrub with a stiff brush.

CLEANING AGENTS

Cleaning agent in general can be defined as natural or synthetic substances that are used to assist the cleaning process-that is, the removal of dirt and dust and the maintenance of a clean appearance on the surface.

WATER

Water is the probably the most common and widely used cleaning agent as it is easily available. It is useful in rinsing and finishing of cleaning processes. Precaution must be taken to change dirty water with fresh water because it could leave a film of dirt instead of removing it.

Warm water dissolves soap more readily than cold water; hence it must be used to remove dirty soap leather. Water is of two types, hard water and soft water.soft water is best for cleaning.

Hardness in water is caused by the presence of mineral salts-mostly calcium (Ca) and magnesium (mg), iron(Fe) and manganese(Mn).the mineral salts reach with soap to form an insoluble precipitate known as soap film or scum.

The result is that the soap film does not rise away easily/it tends to remain behind and produces visible deposits on clothing making it stiff. It attaches to the insides of bathtubs and sinks. Caution must be taken that water is soft as no detergent is effective with hard water.Moreover, hard water does not wet the surface well as a precondition for good cleaning action.

DETERGENTS

These are the cleaning agents that, when used in conjunction with water loosen and remove dirt and then hold it in suspension so that the dirt is not re-deposited on the cleaned surface. They can be of two types. Soapy detergents and synthetic detergents.

The three basic properties of a good detergent are:

- Good wetting power-to lower the surface tension of water and enable the surface of the articles to be thoroughly wet.
- Good emulsifying power-to break up the grease and enable the dirt to be loosened.
- Good suspending power-to suspend the dirt in solution, thus preventing its re-deposition.
- Readily dissolve in water
- Cleanse quickly wit minimum agitation.
- Be effective in all rabges of hard water, without producing scum.
- Be effective over a wide range of temperatures.
- Be harmless to the skin and surface to be cleaned.
- Be easy to rinse away and be biodegerable.

THE CORRECT USE OF DETERGENTS

- Dilute as per the manufacturer's recommendations, using the measuring scoops use the right detergent for the surface to be cleaned.
- Use protective gloves when using strong detergents.
- Dissolve the detergent thoroughly before use.
- Rinse away all traces of detergent from the surface and any cleaning equipment employed.
- Label detergent containers neatly.
- Store detergent containers in a dry, well-ventilated storage area.
- Wipe up any spilled detergent, as it may be a safety hazard.

ABRASIVES

These are substances or chemicals that depend on their rubbing or scratching action to clean dirt and dust from the hard surfaces. They are used to remove very stubborn stains on various surfaces.

TYPES OF ABRASIVES

Based on the scale of hardness for various substances abrasives are classified as;

1	2	3	4	5	6	7	8	9	10
Talc	Calcite			Foldspar				Diamond	

Fig: the scale of hardness of abrasives

Fine Abrasives: these include precipitated whiting (filtered chalk) and jeweler's rouge (a pink oxide of iron) used for shining silver. They are also constituents of commercial silver polishes. Medium abrasives: these include rotten stone, salt, scouring powder and scouring paste. Scouring powders are made up of fine particles of pumice mixed with soap/detergent, and alkali and a little bleach.

Hard/coarse abrasives: these include bath bricks, sandpaper, pumice, steel wool, and emery paper. Glass paper, calcite, sandpaper, fine ash, emery powder and paper, jeweler's rouge, powdered pumice, precipitated whiting (filtered chalk). Ground limestone, sand, steel wool and nylon scourers are some commonly used abrasives. Abrasives are usually not used alone in cleaning agents. For example, a cream or paste meant for cleaning utensils contain about 80% of finely ground limestone, along with other substances such as bleaches, anionic surfactants, alkaline builders, and perfumes.

REAGENTS

These bring about cleaning be a chemical reaction requiring a distinctly low or high pH.pH is a measurement of the level of acid or alkali in solution or substance. In the pH range of 0 to 14 a reading below 7 shows an acid and one above 7 shows an alkali. A pH

Table. pH scale.

1	2	3	4	5	6	7	8	9	10	11	12	13	14
Acidic Neutral Alkaline													

TYPES OF REAGENTS

Acids

Acids used as cleaning agents may vary from mild acids with a pH of 3 to strong acids with a pH of 1.mildly acidic substances used commonly in cleaning include lime, vinegar, tamarind and buttermilk. acids dissolves metal stains, such as water stains in bath tub, hard water around taps and tarnish o silver, copper and brass articles.

- Acids may be used in solution alone or may be part of some special formulations as in toilet cleaners.
- Housekeeping staff need to be trained in the safe handling of strong acids, as they are highly corrosive. They literally'eat away' dirt.
- Rubber gloves should be used while handling them. They should be used in very small quantities as they emit toxic fumes as well.
- Strong acids should be thoroughly rinsed away after the cleaning process.
- Some common acids are: Dilute HCL, Acetic acid, Concentrated HCL, Sodium acid sulphate.
- Acids and their use in cleaning

Acid	pH	Uses
Concentrated HCL	1	Removing stubborn hard-water deposits.
Dilute HCL	1	Removing stubborn scales and deposits from sanitary ware. Removing excess cement from newly cemented tiled areas.
Oxalic	2	Removing stubborn hard-water deposits
Acetic acid	3	Removing tarnish and stains from metals such as copper and brass. Neutralizing alkalis are used in cleaning for preventing colours from running during washing.
Sodium and sulphate	5	Removing hard-water deposits and scales from toilets.

Alkalis

These are used as cleaning agents in the form of liquids and powders. They are particularly useful in the laundry. Very strong alkalis should be used with utmost caution as they are corrosive and toxic. These are called caustic alkalis. Many alkalis at as bleaches. Caustic soda-based cleaning agents are used to clear blocked drains and to clean ovens and other industrial equipment. Alkalis and their use

Alkalis	pH	Uses
Sodium hydroxide(caustic ovens and equipments.	14	Removing stubborn grease from soda
Ammonia	11	Removing stubborn grease
Sodium carbonate detergents.	10	Used as an alkalis builder in synthetic and soapy Clearing blocked drains.
Sodium Perborate	10	Removing stains and whitening due to bleaching action at higher temperature (above 40°C)
Sodium Hypocholorite	9	Removing stains and whitening due to bleaching action on various types of surfaces. Acts as disinfectant.
Sodium Bicorbonate	8	Removing stubborn grease from smooth, delicate surfaces. Removing stains such s tea, coffee and fruit juice.
Sodium Pyroborate (borax)	8	Removing stubborn grease from smooth, delicate surfaces. Removing stains such s tea, coffee and fruit juice.
Sodium Thiosulphite	7	Removing iodine stains.

ORGANIC SOLVENTS

Grease is soluble in organic solvents such as carbon tetrachloride, acetone, turpentine and methylated spirit. Thus, these organic solvents are used extensively in the removal of grease, dry-cleaning of fabrics and stain removal.Solvents are also useful in cleaning surfaces that may be harmed by water. Organic solvents evaporates rapidly from a surface and are therefore ideal for cleaning glass surfaces such s mirrors and windows. Organic solvents should be handled with care as they are harmful to the skin, flammable and poisonous.

DISINFECTANT S AND BLEACHES

Disinfectants aid in the cleaning process by bringing about varying ranges of microbial control. The term' disinfectant' is now used as a general term that covers all kinds of agents that bring about germ control. Most disinfectants have a strong smell and therefore should be used only in recommended amounts in areas where germ control is required.

TYPES OF DISINFECTANTS

Disinfectants can be categorized in terms of their chemical action and compositions

Phenols

These are hydroxyl derivatives of the aromatic hydrocarbon benzene. They are used in dilute or high concentrations to disinfect surfaces in hospitals especially.in hotels, diluted phenols are used with their sharp smell masked by other additives.

Halogens

The elements chlorine and iodine may be used as disinfectants.Chlorine is used both as bleach and as a disinfectant on many surfaces.Iodine is not often used to disinfectant surfaces because it tends to leave brown stains.

Quaternary Ammonium Compounds

These are cationic surfactants useful as bacteriocides. Natural pine oils: pine oils are obtained from pine trees. They are germicidal to some extent, but are mainly added changing formulation for their pleasant smell.

Guidelines for using disinfectants:

- Clean the surface first with detergent and rinse with soft water only
- Rinse away the detergent solution thoroughly before using disinfectants on the surface
- Use the correct disinfectant for the range of disinfection required.
- Use the disinfectant at the correct dilution for it to be effective. Different surfaces may require different dilutions of the disinfectants to act.
- Allow the recommended time for the disinfectant to act.

Glass Cleaners

These are composed of organic, water-miscible solvent such as isopropyl alcohol and an alkaline detergent. Some glass cleaners also contain a fine, mid abrasive. Most glass cleaners are available as sprays or liquid. They are sprayed directly onto windows, mirrors and other glass surfaces or applied on with a soft cloth and rubbed off using a soft, lint-free duster. A glass cloth ideal for the purpose.soft water to which some mathylated spirit or vinegar is added is an inexpensive glass cleaner that can be readily made in the housekeeping department.

Deodorizers

Deodorizers aid in the cleaning process by concentrating stale odours and sometimes also introducing a fragrance to mask them. They are used in restrooms, guestrooms, guest bathrooms, cloakrooms and public areas such as lobbies. Some deodorizers leave no trace of a perfume cover-up. They are usually available as aerosol sprays.

Liquids, powders and crystalline blocks. the crystalline blocks are effervescent and manufactured using the principle of time0 released aromatic chemicals. Naphthalene ball also serve as effective deodorize the crystalline blocks are effervescent and manufactured using the principle of time0 released aromatic chemicals. Naphthalene ball also serve as effective deodorizers. If through cleaning and good ventilation are provided, money need not to be spent on expensive deodorants.

LAUNDRY AIDS

Laundry'aids' are the materials used to improve laundering results(bleaches, optical whiteners) or to accomplish specific functions or effects (soaks, stain removers, softeners, stiffeners. the important laundry agents or acids are water, laundry soaps and detergents, stiffening agents, bleaches, alkaline agents, acid agents, organic solvents, and absorbents.

TOILET CLEANERS

These are strong, concentrated cleaning agents designed to clean and disinfect WCs and urinals. They are available in liquid, powder and crystalline forms. They are acidic in nature as their main function is to remove stubborn stains and limescale.

They all contains some form of disinfectant:

- *Liquid toilet cleaners*: These contain dilute hydrochloric acid, bleach and pine oil. Adequate protective gear should be worn by the user while using such a toilet cleaner.
- *Crystalline toilet cleaners*: These contain sodium acid sulphate, anionic surfactants, and pine oil.
- *Powdered toilet cleaners*: These cleaners contain a soluble acidic powder, chlorinated bleach, a fine abrasive and an effervescing agent to help the active ingredient spread in water.

Whatever the from of toilet cleaner used, they should never be mixed with other cleaning agents since harmful gases may be produced in the resulting reaction.

POLISHES

These chemicals produce a shine by providing a smooth surface from which light is reflected evenly. Polishes are primarily applied to a surface to form a hard, protective layer and thus guard against finger marks, stains and scratches. They also create attractive sheen on the hard surfaces.

CLASSIFICATION OF POLISHES

Polishes are used on metal, furniture and floorings and are classified just as to the type of surface they are used on. On metals they also smooth out any unevenness on the surface of the articles and in case of flooring and furniture they provide a smooth protective layer.

Metal Polishes

These remove the superficial tarnish that forms on metal surfaces due to the attack of certain compounds in the air and some foodstuffs. These polishes also eliminate May scratches on the metal. They consist mainly of a very fine, mildabrasive.in hotels; hard surfaces where metal polishes are used include door-plates and handles, foot rails in bars, staircase banisters, ashtrays, bathroom fittings, tableware and cutlery. In metal polishes such as Brasso, and Silvo, the grease solvent or acid mixed with the abrasive powder aids in the removal of tarnish.

Certain recommendations on the use pf metal polishes are given below:

- Cover the surroundings area with newspaper or protective sheets.
- Carry out any polishing work in a well-ventilated room.
- Use an appropriate polish applicator or disposable rags.
- Use a cocktail stick covered with a rag to apply polish in narrow, hard-to reach nooks and corners.
- Remove polish with cotton or soft, dry cloth.
- Buff the polish with rags and dusters.
- Wash any polish cutlery in warm detergent solution before use.
- Leave the area and equipment clean and tidy after polishing work is complete.
- Dispose of the rags and newspaper used.

Furniture Polishes

These contain a wax or resin, solvent water and a silicone. The main role of wax is to provide a smooth surface from which light is reflected evenly, producing an attractive shine the types of waxes commonly used are carnauba, beeswax, ozokerite, paraffin wax.

Types of furniture polishes:

- *Paste polishes*: these are ideal for use on antique wooden furniture.
- *Cream polishes*: they are used on furniture to give a gloss finish
- *Liquid polishes*: they are used on glossy furniture to remove grease marks and other stains.
- *Spray-on polishes*: they are ideal for use on non-porous surfaces such as glass, chromium, plastic, and varnished or gloss-painted wood.

Considerations for the use of furniture polishes are as listed below:

- Apply the polish on a clean surface.
- Use the least quantity required to accomplish a good polishing of the surface, else the solvent will evaporate and the polish will dry out.
- Always keep the polish container closed when not in use, else the solvent will evaporate and the polish will dry out.
- Use soft, disposable rags, for applying polish to the surface, except in the case of spray-on-polish.
- Use polishes undiluted, unless it is specified otherwise by the manufacturer.
- Be careful while using polishes with a high solvent content since they are flammable.

Floor Polishes

These have a two-folded function. They not only lens an attractive sheen to the surface, but also provide a protective coat on it. The right kind of polish should be used along with the right equipment.

Types of Floor Polishes

Spirit/Solvent-based Polishes

These polishes are used on the porous floor such as wood, wood composition, cork, magnetite and linoleum.

Water-based Polishes

Theses polishes are used on semi-porous surfaces such as thermoplastics, PVC, rubber, asphalt, terrazzo, marble, and natural materials such as Cuddapah tiles and so on.

Considerations for the use of furniture polishes are as listed below:

- Use an appropriate sign to warn people walking along that area of the fact that floor polishing is being carried out.
- ventilate the area well before starting
- Apply the polish to clean, dry floor.
- Rinse the floor thoroughly using a neutralizing agent such as diluted vinegar after stripping the old polish.
- Apply several thin coats of polish rather that a few thick coats.
- Work systematically to ensure that all areas are covered.
- Allow sufficient drying time before applying a second coating.
- Buff thoroughly to reduce the slipping hazard.
- Remove any extra build-up of polish with an appropriate abrasive pad.
- Leave all polishing equipment clean and store them properly.

Leather Polishes

Theses contain a special blend of waxes, a spirit solvent, and occasionally a dye. They are available in the form of creams and liquids. They help to keep the leather supple and impart a shine to it.they also prevent deterioration of old leather articles.

FLOOR SEALERS

These are applied to flooring surfaces as a semi-permanent finish that acts as a protective barrier by preventing the entry of dirt, gems and liquid, grease, stains and bacteria. They prevent scratching and provide an easily maintainable surface. The right type of seal should be applied to each type of floor for effective protection and an attractive appearance. Their functions, floor sealers can be finishing protective or combination of both.

TYPES OF FLOOR SEALERS

- *Oleo-resinous sealers*: they are used for imparting an attractive surface gloss penetrating the floor darkening the colour and highlighting the grain of wood floors.
- *One-pot plastic sealers*: they are used on wood, wood-composition, cork and magnesite floors.
- *Two-pot plastic sealers*: they are also used on wood, wood-composition, cork and magnesite floors.
- *Pigmented sealers*: these may be used on concrete, wood, wood-composition, magnesite, asphalt, and stone floors.

Water-based seals: these may used on marble, terrazzo, magnesite, linoleum, rubber, thermoplastic tiles, PVCs, asphalt, concrete, stone-, and quarry tiles.

Considerations for the use of furniture polishes are as listed below:

- Use appropriate signs to warn passers-by that sealing is being carried out; else it may be a safety hazard.
- Ensure the floor surface is clean, chemically neutral, and dry before applying the sealer. otherwise the seal will not'key' to the floor surface.
- Maintain an ideal room temperature of 21.C
- Keep the room well ventilated.
- Protect the area from flies and pests until the sealer is dry.
- Keep on hand only the required amount of sealer and store the rest tightly lidded, else the whole bulk may deteriorate.
- Apply several thin coats rather than a few thick ones.
- Allow the recommended drying time between coats.
- Clean and store all equipment, such as sealer applicators, neatly after use.

FLOOR STIPPERS

These are uses to remove a worn-out floor finish so that a new sealer or polish can be applied.most are based on alkalis with a high pH.there are two main types available. One is based on ammonia and the other is non-ammoniated product. Ammonia has an intense odour and therefore the area should be well ventilated for many days to get rid of the smell.

CARPET CLEANERS

These are composed of neutral water-soluble solvents, emulsifiers, de-foamers, sanitizers, optical brighteners and deodorizers. They are available as sprays, powders, foams, and liquid shampoos. Whichever they selected it is essential that they are used in correct dilutions.

Some common cleaning agents used in professional housekeeping are:

- *Ammonia*: Liquid ammonia is a solution of ammonia gas in water, held as ammonium hydrxide.it is a strong alkali used for softening water, cleaning window panes, and emulsifying grease.

- *Bath brick*: This is a reddish-brown powder, also obtained in a brick form. it is used fro scouring and polishing metals such as Brass and Copper. In powdered form, it is used for cleaning earthenware.
- *Benzene*: obtained from the distillation of coat tar, benzene is used as a grease solvent and for removing paint and tar stains.
- *Borax*: chemically sodium borate, this white crystalline powder is used to soften hard water to remove coffee and tea stains.
- *Bran*: the husk of wheat grain. It is used in dry-cleaning as grease absorbent.
- *Fuller's earth*: this is ash-white clay that readily absorbs grease. it is used on coloured wood surfaces.
- *Hydrochloric acid*: this is a corrosive and poisonous mineral acid, used diluted for removing stains in bathrooms.
- *Jeweler's rouge*: chemically this is fabric tetroxide, a pinkish powder used for polishing silver. It is a constituent of commercial silver polishes too.
- *Lemon*: lemon is used for removing ink stains from wooden surfaces.
- *Linseed oil*: this is obtained from the crushed seeds of the flax plant. It is a constituent of furniture polishes and paints. It darkens unpainted wood slightly.
- *Magnesia*: chemically magnesium carbonate, this fine white powder is used for dry-cleaning felt, fur and woolen articles.
- *Methylated spirits*: this is used for cleaning window panes and mirrors to a shine. It is a constituent of varnishes and lacquers.
- *Oxalic acid*: this is an organic acid used for the removal of stains from fabrics and bath fittings. It is also used for cleaning porcelain.
- *Paraffin oil*: this is a liquid product of distillation of crude petroleum and is used for cleaning greasy iron and steel articles. It also cleans greasy earthenware when used in combinations with bath brick.
- *Petrol*: This too is obtained from petroleum distillation. It is highly inflammable and is use for dry-cleaning and for removing grease stains.
- *Pumice*: this is a light; porous rock of volcanic origin. It is used as an abrasive for hard metals, earthenware and enamel.
- *Rottenstone*: this is a decomposed siliceous limestone and is used for cleaning copper, brass and earthenware.
- *Common salt*: chemically sodium chloride, this is used as a medium-grade abrasive. It is used for stiffening the bristles of brushes and stiff brooms. Salt is added as a mordant while washing coloured clothes. (A mordant is a substance that prevents undue loss of colour while washing clothes.
- *Sand*: this hard compound of silica is used as a hard abrasive on stone floors and hard, coarse wood.

- *Sikakai*: sometimes called soap-nut or soapberry, but more accurately soap-pod (to distinguish from the fruit of reetha).this is used for non-abrasive cleaning of tarnished metals.
- *Soda*: It emulsifies grease and aids in the cleaning of dirty pans.
- *Steel wool*: this is steel manufactured into long filaments, in varying grades of fitness it is used for scouring hard metals and dirty pans.
- *Turpentine*: this is a constituent of paints.it is also a diluent for paints and removes tar stains.
- *Vaseline*: this is obtained as a residue in petroleum distillation.it prevents rust formation on metals, act as a lubricant, and may be applied on leather to make it soft and supple.
- *Vinegar*: chemically this is 4% acetic acid. It is used to remove stains and tarnish from metals such as copper. It is also effective in removing streaks from glass surfaces such as window panes and mirrors.
- *Whiting/precipitated whiting*: chemically this is calcium carbonate.in pure form. It is used as a mild abrasive on soft metals and in cleaning white-painted articles.

SELECTION OF CLEANING AGENTS

The use of cleaning agent is meant to save time, effort and money. if selected well, all the three objective may be fulfilled.

The following points need to be considered when selecting cleaning agents:

- The type of spoilage
- The type of surface
- Odour
- Range of action or versatility
- Composition of the cleaning agents
- Ease of use, saving of effort and time.
- Toxicity or side-effects.
- Shelf life
- Packing volume and quantities.
- Cost effectiveness.

STORAGE OF CLEANING AGENTS

Cleaning agents with a longer shelf life are usually bought in bulk because of the reduced costs that accrue from the economics of scale. other agents are bought and replenished periodically.storage of cleaning agents is crucial and the various points to be kept in mind.

The points are listed below:

- Ensure that the storage racks are strong and with selves. Heavier containers must be kept on the bottom shelf.
- The store should be kept clean and well-ventilated at all times.

- Ensure that the lids are tightly fitted.
- When issuing cleaning agents use appropriate dispensers and measuring apparatus.
- Ensure that no residual deposits of cleaning agent are left around the rims of the containers.
- Avoid spillage, if a spill occurs, clean it up immediately.
- Follow a systematic procedure for rotating stocks.
- Organic solvents, strong reagents, polishes should be kept away from heat sources.
- Check stock regularly the store should be locked when not in use.

DISTRIBUTION OF CLEANING AGENTS

The housekeeper should implement proper system for the methodical issuing of cleaning agents from the housekeeping stores.

REQUISITIONING

This system of issuing is followed in large hotels. The floor supervisor maintains a requisition book with requisition slips in triplicate. a requisition slip is filled out by a GRA whenever supplies are diminishing. This is signed by the floor supervisor and the book is then sent to the housekeeping stores.

The store keeper collects the requisitioned items and signs the triplicate copies of the requisition slip. The storekeeper then issues the requisitioned items which are collected by a porter and transported to the floor to the room attendant and they should sign in the requisition slip after receiving it.thesecond copy is sent to the executive housekeeper and the third copy remains in the requisition book. Which too is returned with the fresh supplies.

FULL OR EMPTY

This system of issuing is followed in smaller hotels. Empty containers of used-up cleaning supplies are taken to the housekeeping stores by individual GRAs.the store assistant then replaces the empty containers whit full ones.

The disadvantage in this system are that it works well only when the housekeeping stores are open round the clock and that constant supervision is required. The topping-up method is an improvement on this system of issuing.

TOPPING-UP

The difference between this method and full for empty method is that here the GRAs approach the housekeeping stores only at a fixed time each week for getting their supplies toped up. An even better system is having GRAs deposit their hand caddied in the housekeeping stores at the end of the shift, so that the store assistant may replenish or top up the cleaning agents and keep them ready for the staff on the next shift.

THE MAID'S CART AND THE SERVICE ROOM

Maid's cart also called as a room attendants trolley, or chamber maid's trolley, this is perhaps the most significance piece equipment in the housekeeping department. it is like a giant tool box, stocked with everything necessary to service a guest room effectively. Most such carts available now are made of metal, but sometimes wooden carts may be in use. The cart should be spacious enough to carry all the supplies needed for a GRA to complete half a day's room assignments. Since the cart is large and may be heavily loaded, it must be easily movable as well. The ideal cart would have fixed wheels at one end and castor-wheels at the other. The cart should be well organized so that the GRA's do not have to waste time in searching for supplies or make frequent trips back to the supply room. Also, if the cart is not stacked neatly, it will look very unsightly. They are more useful than hand caddies when a large amount of supplies and items are to be carted or replaced. They are ideal for the efficient removal and carriage of smaller pieces of cleaning equipment, cleaning agents, linen and rubbish. They eliminate the time wasted in assembling equipment at the work location or moving them from one place to another.

LOCATION OF THE MAID'S ROOM

Maid's service room/floor pantry: these are located on each guest floor to keep a stock of linen, guest supplies and maid's cart and cleaning supplies for that particular floor. Maid's service room stores a complete set of linen for the whole floor over and above what is already in circulation in the rooms. The maid's service room should be tucked away from guest's view and should be situated near the service elevators. It should store all housekeeping items so that the housekeeping staffs do not have to keep going back to the housekeeping department or linen room for any item. It should have shelves and cupboards door linen and supplies, and sufficient area to park a room maid's cart. It should have a sink with water supply. Since the maid's service room is used to stock expensive items such as linen, it should remain locked at all times when not in use. The key to the floor pantry is kept by the GRA of that floor and a duplicate is kept with the floor supervisor.

The following should be provided in the maid's service room:

- Cupboards to store guest supplies, cleaning agents and equipments
- Shelves and racks to store fresh room linen
- Linen trolleys to store fresh and soiled linen and for transporting/ dispatching the same to the linen and uniform room.
- A notice board to display information regarding expected arrivals, VIPs in house, extra bed, etc.
- A sink with hot and cold water facilities to wash or disinfect glasses, fill drinking water in flasks, and for flower arrangements
- Guest loan items such as rollaway beds, bed boards, baby sitters etc.

CARE AND MAINTENANCE

- All carts and trolleys need to be kept clean, wiped daily and stored locked.
- The room should be dry area and it should be well ventilated.
- The soiled linen bag and thrash bags should be emptied.
- A thorough cleaning may be done once a week.
- The wheels may be oiled during this cleaning for the smooth movement of the trolley.
- Carts and trolleys should be never used as general dumping grounds when not in use.
- All the linens and amenities should be well arranged when the shift is completed so that it will be easier to replenish extra items when necessary.

4

Uses of Clothing

It is not certain when people first started wearing clothes however, anthropologists give estimates that range between 100,000 to 500,000 years ago. The first clothes were made from natural elements: animal skin and furs, grasses and leaves, and bones and shells. Clothing was often draped or tied however, simple needles made out of animal bone provide evidence of sewn leather and fur garments from at least 30,000 years ago.

When settled neolithic cultures discovered the advantages of woven fibers over animal hides, the making of cloth, drawing on basketry techniques, emerged as one of humankind's fundamental technologies. Hand and hand with the history of clothing goes the history of textiles. Humans had to invent weaving, spinning and other techniques and the machines needed in order to make the fabrics used for clothing.

Clothing is defined, in its broadest sense, as coverings for the torso and limbs as well as coverings for the hands (gloves), feet (socks, shoes, sandals, boots) and head (hats, caps). Humans nearly universally wear clothing, which is also known as dress, garments, attire, or apparel. People wear clothing for functional as well as for social reasons. Clothing protects the vulnerable nude human body from the extremes of weather, other features of our environment, and for safety reasons. Every article of clothing also carries a cultural and social meaning. Human beings are the only mammals known to wear clothing, with the exception of human pets clothed by their owners.

People also decorate their bodies with makeup or cosmetics, perfume, and other ornamentation; they also cut, dye, and arrange the hair of their heads, faces, and bodies (see hairstyle), and sometimes also mark their skin (by tattoos, scarifications, and piercings). All these decorations contribute to the overall effect and message of clothing, but do not constitute clothing per se.

Articles carried rather than worn (such as purses, canes, and umbrellas) are normally counted as fashion accessories rather than as clothing. Jewelry and eyeglasses are usually counted as accessories as well, even though in common speech these items are described as being worn rather than carried.

CLOTHING AS FUNCTIONAL TECHNOLOGY

The practical function of clothing is to protect the human body from dangers in the environment: weather (strong sunlight, extreme heat or cold, and precipitation, for example), insects, noxious chemicals, weapons, and contact with abrasive substances, and other hazards. Clothing can protect against many things that might injure the naked human body. In some cases clothing protects the environment from the clothing wearer as well (example: medical scrubs).

Humans have shown extreme inventiveness in devising clothing solutions to practical problems and the distinction between clothing and other protective equipment is not always clear-cut. See, among others: air conditioned clothing, armor, diving suit, swimsuit, bee-keeper's costume, motorcycle leathers, high-visibility clothing, and protective clothing.

Clothing as Social Message

Social messages sent by clothing, accessories, and decorations can involve social status, occupation, ethnic and religious affiliation, marital status and sexual availability, etc. Humans must know the code in order to recognize the message transmitted. If different groups read the same item of clothing or decoration with different meanings, the wearer may provoke unanticipated responses.

Occupation

Military, police, and firefighters usually wear uniforms, as do workers in many industries. School-children often wear school uniforms, while college and university students sometimes wear academic dress. Members of religious groups may wear uniforms known as habits. Sometimes a single item of clothing or a single accessory can declare one's occupation or rank within a profession — for example, the high toque or chef's hat worn by a chief cook.

Ethnic, Political, and Religious Affiliation

In many regions of the world, national costumes and styles in clothing and ornament declare membership in a certain village, caste, religion, etc. A Scotsman declares his clan with his tartan. A Muslim woman might wear a hijab to express her religion. A male Sikh may display his religious affiliation by wearing a turban and other traditional clothing. A French peasant woman may identify her village with her cap or coif.

Clothes can also proclaim dissent from cultural norms and mainstream beliefs, as well as personal independence. In 19th-century Europe, artists and writers lived la vie de Bohème and dressed to shock: George Sand in men's clothing, female emancipationists in bloomers, male artists in velvet waistcoats and gaudy neckcloths. Bohemians, beatniks, hippies, goths, punks and Skinheads have continued the (counter-cultural) tradition in the 20th-century West. Now that haute couture plagiarises street fashion within a year or so, street fashion may have lost some of its power to shock, but it still motivates millions trying to look hip and cool.

Marital Status

Hindu women, once married, wear sindoor, a red powder, in the parting of their hair; if widowed, they abandon sindoor and jewelry and wear simple white clothing. Men and women of the Western world may wear wedding rings to indicate their marital status. See also Visual markers of marital status.

Sexual Availability

Clothing may signal an individual's receptiveness to sexual advances. Some garments signal lack of interest in advances; some garments and accessories indicate openness to flirtation. What constitutes modesty and allurement varies radically from culture to culture, within different contexts in the same culture, and over time as different fashions rise and fall. Often, exposure of skin and hair is an availability signal; covering skin and hair signals unavailability. However, minute adjustments of "modesty" signals can subvert the surface meaning and convey a mixed message ("I'm nice but I like to flirt too).

The vocabulary of women's clothing is usually more developed than the vocabulary of men's clothing in this regard.

Examples of sexual signaling:

- In Amish communities, both men and women dress in plain garments that cover the body, without intricate details or patterns. Women also wear a prayer veil. Unmarried women wear black veils, married women wear white ones.
- Many Muslim women wear a head or body covering (see hijab, burqa or bourqa, chador and abaya) that proclaims their status as "respectable" and modest women. Individual women choose to wear these coverings in vastly different, and to some, seemingly conflicted ways. One woman may wear a very plain, drab, and loose-fitting abaya, while another woman may choose an abaya of luxurious material cut close to the body and then accessorize with high heels and a fashionable purse.
- Streetwalking prostitutes in countries such as the United States where prostitution is illegal must dress to advertise their status to potential customers, while avoiding anything that might constitute an unambiguous offer of sex for sale (which would increase their chances of being caught and convicted). They tend to wear current fashions in exaggerated form, bare a great deal of skin, and wear heavy makeup.
- An American or European woman who wants to signal availability must sport some culturally accepted signals of flirtatious intent, but

without the exaggeration that might lead others to say that she is dressing like a prostitute. In the last few decades, there has been a consistent trend towards the mainstreaming of formerly extreme fashion, in which "over-the-top" becomes ordinary and loses any shock value it might once have had.

HISTORY OF CLOTHING

According to archaeologists and anthropologists, the earliest clothing probably consisted of fur, leather, leaves or grass, draped, wrapped or tied about the body for protection from the elements. Knowledge of such clothing remains inferential, since clothing materials deteriorate quickly compared to stone, bone, shell and metal artifacts. Archeologists have identified very early sewing needles of bone and ivory from about 30,000 BC, found near Kostenki, Russia, in 1988.

Ralf Kittler, Manfred Kayser and Mark Stoneking, anthropologists at the Max Planck Institute for Evolutionary Anthropology, have conducted a genetic analysis of human body lice that indicates that they originated about 107,000 years ago. Since most humans have very sparse body hair, body lice require clothing to survive, so this suggests a surprisingly recent date for the invention of clothing. Its invention may have coincided with the spread of modern Homo sapiens from the warm climate of Africa, thought to have begun between 50,000 and 100,000 years ago. However, a second group of researchers used similar genetic methods to estimate that body lice originated about 540,000 years ago. For now, the date of the origin of clothing remains unresolved.

Some human cultures, such as the various peoples of the Arctic Circle, until recently made their clothing entirely of furs and skins, cutting clothing to fit and decorating lavishly.

Other cultures have supplemented or replaced leather and skins with cloth: woven, knitted, or twined from various animal and vegetable fibres. See weaving, knitting, and twining.

Although modern consumers take clothing for granted, making the fabrics that go into clothing is not easy. One sign of this is that the textile industry was the first to be mechanized during the Industrial Revolution; before the invention of the powered loom, textile production was a tedious and labor-intensive process. Therefore, methods were developed for making most efficient use of textiles.

One approach simply involves draping the cloth. Many peoples wore, and still wear, garments consisting of rectangles of cloth wrapped to fit — for example, the dhoti for men and the saree for women in the Indian subcontinent, the Scottish kilt or the Javanese sarong. The clothes may simply be tied up, as is the case of the first two garments; or pins or belts hold the garments in place, as in the case of the latter two. The precious cloth remains uncut, and people of various sizes or the same person at different sizes can wear the garment.

Another approach involves cutting and sewing the cloth, but using every bit of the cloth rectangle in constructing the clothing. The tailor may cut triangular pieces from one corner of the cloth, and then add them elsewhere as gussets. Traditional European patterns for men's shirts and women's chemises take this approach.

Modern European fashion treats cloth much more prodigally, typically cutting in such a way as to leave various odd-shaped cloth remnants. Industrial sewing operations sell these as waste; home sewers may turn them into quilts.

In the thousands of years that humans have spent constructing clothing, they have created an astonishing array of styles, many of which we can reconstruct from surviving garments, photos, paintings, mosaics, etc., as well as from written descriptions. Costume history serves as a source of inspiration to current fashion designers, as well as a topic of professional interest to costumers constructing for plays, films, television, and historical reenactment.

Future Trends

As technologies change, so will clothing. Many people, including futurologists have extrapolated current trends and made the following predictions:

- Man-made fibers such as nylon, polyester, terylene, terycot, lycra, and Gore-Tex already account for much of the clothing market. Many more types of fibers will certainly be developed, possibly using nanotechnology. For example, military uniforms may stiffen when hit by bullets, filter out poisonous chemicals, and treat wounds.
- "Smart" clothing will incorporate electronics. Clothing may incorporate wearable computers, flexible wearable displays (possibly leading to fully animated clothing and some forms of invisibility cloaks), medical sensors, etc.
- Present-day ready-to-wear technologies will presumably give way to computer-aided custom manufacturing. Low power laser beams will measure the customer; computers will draw up a custom pattern and execute it in the customer's choice of cloth as well as choice of desirable fit.

CLOTHING FETISHES

Because clothing and adornment have such frequent links with sexual display, humans may develop clothing fetishes. They may strongly prefer to have sexual relations with other humans wearing clothing and accessories they consider arousing or sexy. In Western culture, such fetishes may include extremely high heels, lace, leather, or military clothing. Other cultures have different fetishes. The men of Heian Japan lusted after women with floor-sweeping hair and layers of silk robes. Fetishes vary as much as fashion.

Sometimes the clothing itself becomes the object of fetish, such as in case with used girl panties in Japan.

Religious Habits and Special Religious Clothing

Religious clothing might be considered a special case of occupational clothing. Sometimes it is worn only during the performance of religious ceremonies. However, it may also be worn everyday as a marker for special religious status.

For example, Jains wear unstitched cloth pieces when performing religious ceremonies. The unstitched cloth signifies unified and complete devotion to the task at hand, with no digression.

The cleanliness of religious dresses in Eastern Religions like Hinduism, Buddhism and Jainism is of paramount importance, which indicates purity.

Sport and Activity

Most sports and physical activities are practised wearing special clothing, for practical, comfort or safety reasons. Common sportswear garments include shorts, T-shirts, polo shirts, tracksuits, and trainers. Specialised garments include wet suits (for swimming, diving or surfing) and salopettes (for skiing).

Clothing Materials

Common clothing materials include:

- Cloth, typically made of cotton, flax, wool, hemp, ramie, silk, lyocell, or synthetic fibers
- Down for down-filled parkas
- Fur
- Leather

Less-common clothing materials include:

- Paper
- Rubber
- Recycled PET
- Rayon
- Bamboo
- Soy
- Jute
- PVC
- Tyvek
- Hemp
- Recycled or Recovered Cotton
- Other Natural Fibers

Reinforcing materials such as wood, bone, plastic and metal may be used in fasteners or to stiffen garments.

CLOTHING CARE AND FIBER CONTENT LABELS

In most families stain removal is a necessary part of clothing care. Quick and cautious stain removal keeps clothes in wearable condition for a longer time and thus helps reduce clothing costs. This bulletin explains how to remove many stains from washable fabrics using easily available laundry supplies.

You know if a fabric is washable by reading the care label. If the label has warnings such as "do not use chlorine bleach," then you cannot safely use liquid chlorine bleach in stain removal.

Clothing manufacturers are required to tell you the recommended care procedure on a "permanent care label." These labels often appear at the neckline of garments, but sometimes they are in a side seam. These labels tell if the manufacturer recommends home laundering or drycleaning for routine care of the clothing. A recommendation for drycleaning may relate to the component parts of the garment such as interfacings and trim, rather than to the basic fiber content of the garment. Tailored clothes of wool are often "dry-clean only" while wool sweaters are often "hand washable."

Hang tags or other labels on clothing tell the fiber content. If you understand fiber characteristics, this information can help you decide about the best stain removal procedure. For example, wool or silk can be severely damaged by liquid chlorine bleach, so this bleach should be avoided. Hang tags may also tell about fiber finishes such as soil release or durable press that are not visible on clothing but can make stain removal easier or more difficult. For example, oily stains bond more firmly with durable press fabrics than with untreated fabrics, making removal more difficult.

Fabrics Labeled "Dry-clean Only"

The procedures described here do not apply to garments labeled dry-clean only. Clothing labeled "dry-clean" or "professionally dry-clean" should be taken to the cleaners promptly. The fiber content of the clothing and the type of stain should be identified for the cleaner. Even professional cleaners cannot remove all stains. The cleaner will usually warn you if the stain cannot be removed, but sometimes this is difficult to predict. Dye stains on colored fabrics are an example of stains that may not be completely removed.

A Note about Modern Fabrics

Synthetic fibers such as acrylic, nylon, olefin, polyester, and blends of these fibers or cottons with permanent-press finishes are tough and durable, but have a special attraction for oil stains. Oil stains should be removed promptly. If oil stains get dryer dried or ironed into fabrics containing these fibers or finishes, removal may be extremely difficult, if not impossible. These oil stains show most on plain medium colors such as light blue or khaki. On the other hand, if treated quickly, oil stains usually can be easily removed.

Synthetic fiber fabrics are also heat sensitive. These fabrics shrink and melt in high temperatures. They can get more-or-less "permanent" wrinkles in the spin-cycle of a washing machine set for a hot-water wash, or from an over-crowded dryer that has run too long. Steam pressing can sometimes remove heat-set wrinkles, but the melting temperature of the fibers is so close to the temperature needed to iron out wrinkles that pressing is tedious and must be done carefully to avoid melting and creating holes. The restored garment may fit differently because of the heat shrinkage.

It is easy to prevent wrinkling in washable blends and permanent press clothes:

- Avoid overloading the washer; clothes should move freely,
- Be sure the washer is set for "warm" not "hot" water temperature or for the "permanent press" cycle,
- Dry on permanent press setting,
- Remove from dryer at end of cycle; do not overdry,
- Hang on hanger; temporary wrinkles generally "elax" or fall out in a few hours.

Fabrics containing -vinyl or natural rubber will be damaged by most oil solvents. Oil solvents tend to remove the plasticizer in vinyl film fabrics, making them stiff. Olefins may be damaged by perchloroethylene solvent, but are resistant to trichloroethylene and fluorocarbon drycleaning solvents.

Acetate fabrics will dissolve in fingernail polish remover (acetone). Triacetate and modacrylic fabrics can be damaged by acetone or paint thinners. Silk, wool, and other hair fibers, such as camel or cashmere, will dissolve in fresh liquid chlorine bleach. Dilute solutions of liquid chlorine bleach will cause permanent yellowing and stiffening of wool fibers and usually cause weakening and color loss in silk.

Cellulosic fibers, such as cotton, linen, rayon, and ramie, will be weakened by repeated exposure to dilute solutions of liquid chlorine bleach, but bleaches can be safely used on cellulosic fibers for purposes of stain removal. Undiluted bleach can weaken fabrics so that they tear or wear out more quickly.

Garments with Contrasting Colors or Trim

Many garments are designed with dark fabrics and white trim or white fabrics and bright colored trim such as red piping. When these fabrics are labeled washable, people often complain that the colors have "run." The white shirt becomes streaked with pink from the red trim.

Sometimes this problem can be resolved by rewashing the garment in a heavy-duty detergent with warm or hot water. The excess dye that was not permanently in the trim is not likely to be very firmly deposited in its new location either, and sometimes a simple repeat washing will rinse it away. Sometimes bleaching will help, but often the bleach will change the color of the trim and further change the appearance of the garment.

Since there are no regulations about color-fastness labeling, there is little that consumers can do when they have problems with color-fastness except complain to the manufacturer or retailer who sold the goods. As long as care labeling procedures have been carefully followed, most reputable businesses will respond to your complaint in a positive way. Although it may cramp your style, a way to avoid these problems is to not buy clothing with extreme color contrasts.

Removing Stains from Washable Fabrics

The following general procedures apply to nearly all stains. Fresh stains are much easier to remove than old ones, so take care of stains promptly.

- Blot up any excess liquid with a clean white cloth, paper, or other towels. Remove excess solids by gentle scraping or chipping with a dull knife or metal spatula. With some solids such as heavy amounts of surface mud removal may be easier after the stain has dried. Excess can be brushed off before the clothing is submerged for washing.
- Avoid rubbing the stained area with a linty terry towel or a dark-colored cloth. You may complicate the problem.
- Never rub a fresh stain with bar soap. Soap sets many stains.
- Decide if the fabric is washable or drycleanable. If drycleanable, take to the cleaners as soon as possible (within 24 to 48 hours).
- Do not try to treat suede, leather, or fur. Professional cleaners are needed for these items, and even some professionals do not offer this service.
- Avoid using hot water on stains of unknown origin. Hot water can set protein stains such as milk, egg, or blood.
- Test stain removal agents on a seam or hidden area of the garment to be sure it does not affect the color or finish of the fabric before starting on the stain.
- Avoid excessive rubbing unless fabric is tough and durable. Rubbing can spread the stain and damage the fiber, finish, or color.
- Do not iron or press stained fabrics. Heat will set most stains.
- Check laundry for stains before washing. Many stains need pretreatment.
- Inspect wet laundry before drying to be sure stain has been removed. If a stain is still evident, do not dryer dry. The heat of drying will tend to make the stain more permanent.
- Wash heavily soiled items separately. During laundering soil is broken into smaller particles and can be redeposited on cleaner clothing if insufficient detergent is used, water temperature is too, low, washing time too long, or washer is overloaded with too many clothes.

Spot Treatment Technique (Sponging) for Apparel Fabrics

A spot treatment confines the stain to a small area and keeps it from spreading. This method is sometimes called "sponging." For spot treatment you need a supply of absorbent material such as, clean rags or paper towels, and a drycleaning solvent, spot remover, or aerosol pretreatment spray.

Follow these steps:

1. Pad the working surface with clean rags or paper towels that can be stained as you work.
2. Place the stained area or spot on the garment face-down over the padded surface.
3. Dampen a small white cloth with solvent.
4. Use the dampened cloth to pat the stain from the wrong side. Feather the edges of the stain working from the outside toward the center to keep the stained area from getting larger.
5. As the stain transfers to the absorbent material beneath, move it to a different place on the absorbent material so the stain has a clean place to exit into.
6. Repeat this procedure until all traces of stain are gone. Launder to remove any ring that might be left by the solvent.

Chemical Solvents and Supplies

Supplies needed for stain removal can usually be found in grocery, drug, general merchandise, or paint stores.

Many commercially available products have proprietary formulas that are protected by patent rights and not available to the public. Ingredients listed on labels for safety purposes can help you decide if the product will be useful for the stain removal task- you havė to do. There is no miracle product that will remove all stains.

Detergents

- Heavy-duty liquid detergents (Era, Fab, Grease Relief, Tide, Wisk)
- Light-duty liquid detergents (Delicare, Ivory, Lux,Woolite)
- Powdered detergents (Amway, Cheer, Dash, Oxydol, Sears, Tide)

Bleaches

- Hydrogen peroxide
- Powdered all-fabricc bleaches (sodium perborate)(Biz, Borateem, Clorox 2; Purex, Snowy)
- Liquid all-fabric bleaches (Snowy, Vivid)
- Liquid chlorine bleach (sodium hypochlorite)(Clorox, Hi-lex,Purex)
- Liquid chlorine bleaches have a limited shelf-life. If your bleach is more than six months old and has no effect on stains, it may need to be replaced with fresh bleach.
- To test for colorfastness to liquid chlorine bleaches, mix I tablespoon of bleach with 1/4 cup of water. Use an eyedropper to put a drop of this solution on a hidden seam or pocket edge inside the garment. Let it stand two minutes, then blot dry. If there is no color change it

is safe to use the product. Powdered bleach packages have directions for doing colorfastness tests.

Pretreatment Products

- Aerosol sprays-petroleum-based solvent (Clorox, Shout, Spray'n Wash)
- Pump-type sprays— detergent based (Clorox, Shout, Spray'n Wash)

Absorbent Materials

- Clean white cloths
- Paper towels (white)
- Sponges (white or neutral colored)
- Caution: Colored sponges or paper towels can bleed dyes onto fabrics, making dye stains that may be difficult to remove.

Odor-reducing Agents

- Activated charcoal
- Calcium carbonate
- Soda

Stain Removal Chemicals

Common Name	*Chemical*	*Brand Names*
Alcohol (rubbing)	Isopropyl	-
Ammonia	ammonium hydroxide	Ajax, Bo Beep, Top Job
Color remover	sodium hydrosulfite	Rit, Tintex
Commercial stain removers*	isopropyl alcohol or other unspecified ingredients	Easy Wash, Tech, Whizz
Drycleaning fluid or petroleum-based pretreatment solvent	perchloroethylene, trichloroethylene petroleum distillates	Energine, Clorox Prewash, Shout, Spray n'Wash
Enzyme presoak products**	amylase, protease, lipase	Biz Bleach, Axion
Lemon juice and salt	citric acid and sodium chloride	-
Nail polish remover	acetone	-
Rust removers***	hydrofluoric acid, oxalic acid	RoVer, Whink
Photo supply acid fixer	sodium thiosulfate	-
Turpentine	terpene	-
White vinegar	acetic acid	-

**Limited testing on these products has shown them to be less effective than advertised.*

***These must be used at body temperature for enzyme action to occur, Chlorine bleach and hot water inactivate enzymes.*

****Do not use these products with chlorine or oxygen type bleaches.*

Note: *Some of these products may no longer be available; while new products have also been developed.*

Follow These Safety Precautions

While many stain removal, chemicals and bleaches are commonly used around the home, they are still dangerous. Some are flammable; others are toxic. Here are some rules to keep in mind regarding their use:

- Store stain removal materials out of the reach of children. Do not store with food products.
- Read all label directions and warnings. Store chemicals in their original containers so label directions are available in case of an accident.
- Use all chemicals and commercial stain removal product according to label recommendations.
- Avoid getting the chemicals on your skin. Some are easily absorbed.
- Use drycleaning solvents and other chemicals in a well-ventilated room. Toxic fumes can cause illness.
- Do not use solvents near an open flame or electrical outlet.
- Seal containers so that fumes can't escape.
- Never mix stain removal materials together (bleach and ammonia together form toxic fumes).

Classification of Stains

The system used here in classifying stains for removal from washable fabrics is not the only one that has been used. Other books or sources may suggest different methods that also may work. The purpose here is to describe at least one method that should give good results using readily available consumer products or supplies if used correctly.

This stain classification system starts with stains that require similar treatment and are easiest to remove if treated promptly and correctly. Stains that require two-step or special treatment are listed last.

Protein Stains

Soak in cold water. Launder.

- Baby food
- Baby formula
- Blood
- Mud
- Pudding
- Urine
- Vomit
- White glue; school paste
- Milk
- Mucous
- Cheese sauce
- Cream
- Egg
- Feces
- Gelatin
- Ice cream

Fresh protein stains can be removed by soaking and agitating in cold water before washing. These stains contain other ingredients besides protein, but it needs treatment first. If hot water is used first, it cooks the protein, causing coagulation between the fibers in the yarns of the fabric, making the stains more difficult to remove. If protein stains are dried or old, scrape or brush off crusted matter (if any), then soak in cold water using a detergent or an enzyme presoak product.

After treating the stain, launder in warm (not hot) water, rinse, and inspect. If stain remains, soak an additional half-hour, then rewash. Bleach may be necessary if the stain was colored, such as baby food beets, strawberry gelatin, or ice cream.

Tannin Stains

Do not use soap (bar, flake). Use detergents.

- Alcoholic beverages
- Berries (cranberries, raspberries, strawberries)
- Felt-tip water color pen or washable ink
- Soft drinks
- Tomato juice
- Beer
- Coffee
- Cologne
- Fruit juice (apple, grape, orange)
- Tea

Fresh tannin stains are usually removed by detergent (not soap) washing in hot water (as safe for fabric) during laundering without any treatment. Use of soap (bar soap, soap flakes, or detergents containing natural soap) will make a tannin stain permanent or at least more difficult to remove. Be sure to check the ingredients list of your detergent for soap. More brands now include it for economic reasons. Old tannin stains may need bleaching for more complete removal.

Oil-based Stains

Use heavy-duty detergent with hot water.

- Automotive oil
- Bacon fat
- Butter/margarine
- Car door grease
- Collar/cuff greasy rings
- Cooking fats and oils
- Face creams
- Hair oil
- Hand lotion
- Lard
- Mayonnaise
- Salad dressing
- Suntan oil or lotion

Oil stains can be removed by pretreatment with a heavy-duty liquid detergent, an aerosol petroleum-based solvent pretreatment spray, or a pump-type detergent-based pretreatment spray. If these products are unavailable, you can use a powdered detergent that is mixed with water to make a runny paste and apply that to the stain. The heavy-duty liquid detergents or aerosol sprays are more convenient and effective. Work the full-strength heavy-duty liquid detergent into the stain or spray with the pretreatment product, then wash the garment using hot water (if safe for fabric), the recommended amount of detergent for a regular laundry load, rinse, and inspect before drying. Repeat this treatment if removal is incomplete the first time.

Dye Stains

Need detergent wash and bleach as safe, for fabric.

- Cherry, blueberry (dye transfer)
- Color bleeding in wash
- Felt-tip pen (permanent ink-may not come out)
- Grass
- India ink
- Kool-Aid
- Mercurochrome
- Mustard
- Tempera paint

Dye stains are very difficult to remove. First,pretreat the stain with a heavy-duty liquid detergent, then rinse thoroughly. Soak the stained garment in a dilute solution of all-fabric powdered bleach. If the stain persists, and the garment is white or colorfast, soak in a dilute solution of liquid chlorine bleach and water. Bleaching damage to colored garments is irreversible. To decide if a fabric can be bleached safely, use the test described previously. If the stain is not removed in 15 minutes, it cannot be removed by bleaching and further bleaching will only weaken the fabric. Caution: Since bleaches can alter the color of a fabric as well as the stain, bleach the whole garment and do not try to bleach just a spot.

Combination Stains

Two step treatment: (1) Remove oily/waxy portion, (2) Remove dye portion using bleach as safe for fabric.Combination stains contain a variety of ingredients, but these stains usually have an oily/waxy component and a dye or pigment component. Use the procedures recommended for removing oil stains first.

Step 1 procedure depends on whether stain is in Group A or B as follows: Group A. Spray or sponge with drycleaning solvent (perchloroethylene, trichloroethylene) then rub with heavy-duty liquid detergent before washing.

- Ball-point ink
- Candle wax
- Carbon paper
- Carbon typewriter ribbon
- Crayon
- Eye make-up (mascara, pencil, liner, shadow)
- Floor wax
- Furniture polish
- Lipstick
- Livestock paint
- Pine resin
- Shoe polish
- Tar

Group B. Rub heavy-duty liquid detergent into stain before washing.

- Barbecue sauce
- Calomine lotion
- Catsup or tomato sauce
- Cocoa or chocolate

- Face make-up (powder, rouge, foundation)
- Gravy
- Hair spray

After you've done the procedures above, do step 2-removing dye stains. Start with an all-fabric bleach because it is less damaging to colors and fabrics. Use liquid chlorine bleaches for tough dye stains, if fabrics are colorfast to bleach.

If you don't know what the stain is, its odor, location, and color may give you a clue. Old oil stains may smell rancid, but appear dry. Food stains are often on the front of garments; perspiration stains around collars and underarms; black grease is often on pants or skirts at car-door latch levels.

Stain color may be a misleading clue. For example, rust-colored stains may be coffee, tea, old lemonade stains (carmelized sugar), cosmetics containing benzoil peroxide (which can bleach many colors to look rusty), felt marker, crayon, aged baby formula, or a number of other things. If a heavy waxy or gummy residue is present, you may be dealing with a stain that will respond best to spot treatment with a drycleaning fluid.

Since the appropriate removal method varies with the stain, start by using the least destructive stain removal methods first. If the whole garment can be submerged, start by soaking the garment in cold water (as for protein stains). If not, use warm water and spot treatment technique. Next, use liquid detergent and lukewarm or hot water, rinse and let air dry (as for oil stains).

If you suspect the stain is iron rust, treat with rust remover before bleach. If stain persists, use a pretreatment spray or solvent (as for combination stain) and all-fabric bleach. If the all-fabric bleach is ineffective on the stain and the garment is colorfast or white, finally try a dilute solution of liquid chlorine bleach.

Stains Needing Unique Treatment Methods

Chewing gum: Apply ice to harden gum. Crack or scrape off excess. Spray with pretreatment aerosol product. Rub with heavy-duty liquid detergent. Rinse with hot water. Repeat if necessary. Launder. Deodorants: Apply liquid detergent,wash in warm water. Build-up of aluminum or zinc salts may be impossible to remove.

Fingernail polish: Do not use nail polish remover (or acetone) on acetate, triacetate, or modacrylic fabrics as they will dissolve. Take these fabrics to professional drycleaners and identify the stain. For other fabrics, use nail polish remover,acetone and spot treatment method. Hog confinement odor: Wash clothes adding 1/4 to 1/2 cup household ammonia to wash load with heavy-duty detergent. Do not mix ammonia and bleach in same wash load. Toxic fumes are produced. Ammonia can be used on colored fabrics, but occasionally its use will change the garment's color.

Iodine: Iodine is quickly removed with sodium thiosulfate, which is sold in photo supply stores as "acid fixer." If the photo supply fixer solution contains other chemicals in addition to sodium thiosulfate, it should not be used. Iodine may also be removed by some commercial stain removers. Lead pencil: Use art gum eraser to lift off excess; avoid hard rubbing. For delicate fabrics use spot treatment methods. For most durable, washable fabrics, spray with pretreatment aerosol product. Rub in heavy-duty liquid detergent. Rinse in warm water. Launder.

Mildew: Mildew is a growing organism that must have warmth, darkness, and moisture to survive. Mildew eats cellulosic fibers, causing permanent damage and weakening of fibers and fabrics. To remove mildew: Shake or brush item outdoors. Pretreat darkest stains with heavy-duty liquid detergent. Launder in hot water with a heavyduty detergent. Bleach as safe for fabric.

Odor: Most odors are removed by laundering.For persistent odor problems, place calcium carbonate crystals, activated charcoal, or soda in an open container and store with clothes in closet or sprinkle soda directly on fabric and let stand; then shake or vacuum. Paint-latex: Treat while wet. Soak in cold water; wash in cool water with heavy-duty detergent. After paint has dried 6 to 8 hours, removal is very difficult. Treat as combination stain. Wash in hot water, Rinse. Repeat treatment.

Paint-oil-based: Treat while wet. Use thinner recommended for paint. Use spot treatment technique and thinner on spots until paint is softened and can be flushed away in heavy-duty detergent wash. Usually turpentine or alcohol will work as solvents. Perspiration: Apply liquid detergent or soak in warm water with presoak product 15 to 30 minutes. Launder.

Pesticide: If full-strength liquid concentrate spills on clothes, handle only with rubber gloves. Discard clothing immediately. Laundering does not remove concentrate to a safe level for reuse of clothing. Launder other pesticide-contaminated clothing separate from general family laundry. If visible staining from diluted spray of pesticide residues remains after laundering, rewash using hot water, heavy-duty detergent, and a full water level. Then line dry.

Rust: Rust stains cannot be removed in normal laundering. Use of chlorine bleach makes them permanent. Rust removers such as RoVer or Whink are effective and safe for most fabrics, but rust removers that contain hydrofluoric acid are extremely toxic, can burn the skin, and can damage the finish on appliances. A solution oxalic acid crystals in water will also remove rust stains,- but it is often difficult to obtain the crystals. Lemon juice and salt are more readily available and are helpful sometimes, Sprinkle the salt on the stain, squeeze lemon juice on it and spread the garment in the sun to dry. A word of caution: Lemon juice can bleach some colors and many washable garments are not manufactured to be colorfast to sunlight.

Scorch: Excess heat on cellulosic (cotton, linen, ramie, rayon), wool, or synthetic fibers can cause permanent damage. If fabric is thick and fuzzy, brush to remove charring. Rub liquid detergent into scorched area. Launder. If stain remains, bleach using,all-fabric bleach. Fabric will be permanently weakened in scorched area. Synthetic blends that are melted or glazed cannot be fully restored. Smoke, soot: Shake off excess soot outdoors. Launder in washing machine using heavy-duty phosphate-based detergent or heavy-duty liquid as recommended by manufacturer, one cup of water conditioner, and 1/2 cup of all-fabric bleach. Use water temperature appropriate for fabric. Air dry. Inspect for smoke odor. Repeat as necessary. Three or four washes may be needed for cottons and cotton blends.

Urine: Rinse in cold water and launder. For stains on mattresses: (1) sponge with cloth using detergent solution, (2) rinse with cloth using vinegar solution, (3) let air dry, and (4) if odor remains, sprinkle with soda or calcium carbonate; wait 1 day, then vacuum. Water Spots: Launder. For drycleanable draperies, consult a professional cleaner. Water marks on drapes are water soluble and not removable by drycleaning solvents.

Common Remedies to Avoid

Dishwasher detergent: Although sometimes suggested for food stains, these detergents are intended for use in closed dishwashers with very hot water. They are so highly alkaline they can irritate your skin if you use them in stain removal. They also may fade colors or damage wool, silk, or nylon fibers. Hair spray on ball-point ink: Certain hair sprays are effective on ballpoint stains, but they may deposit a gummy residue and perfume that then have to be removed along with the ink. Hair spray also may affect color in some fabrics. Alcohol is a hair spray ingredient that is useful for removal of the oily part of the ball-point stain.

Ironing candle wax: Ironing candle wax between blotting paper will only drive the stain deeper into the fabric. This process is widely used, but it's not recommended because it will make any color from the dye of the candle more permanently set and the wax more inaccessible for the detergent or solvent to reach to carry the stain away. Milk on washable ink: This doesn't remove the ink and gives you an additional protein stain.

Salt to make dyes colorfast: Today's dyes cannot be increased in colorfastness by soaking in salt water. If bleeding of a particular dye in cotton, rayon, or ramie fabric is decreased with a salt water soak, the effect will not be permanent, When the fabric is wet again, unless there is salt in the solution, the dye will be free to leave the fabric. Salt cannot affect colorfastness of synthetic fiber fabrics or their blends because they are dyed with dyes that have chemical structures not affected by salt. Shampoo: Clear gel-like shampoos are sometimes suggested for stain removal. While they are usually not harmful to fabrics and may work on light oil stains, laundry detergents are just as effective and less expensive to use. Additionally, colored, opaque; or milky-looking shampoos may contain ingredients that will stain fabrics or foam so much that they are difficult to rinse out.

White vinegar: Vinegar (acetic acid) may weaken cotton, rayon, acetate, triacetate, or silk fibers and may cause color change. If used as a stain removal agent, test on a hidden seam allowance for colorfastness. Vinegar will not help remove or set creases in today's synthetic or permanent press fabrics, although this is a common belief.

Identify and Prevent Some Common Staining Problems

Greasy-looking fabric softener splotches: Use of fabric softener sheets in the dryer can deposit softener unevenly, causing greasy-looking splotchy stains on silk-like polyester and blends of cotton/polyester broadcloth. This problem is especially noticeable on medium-colored fabrics such as khaki and medium blue. Avoid this problem and control static by using a fabric softener that is added to the final rinse.

Odd colored or rusty looking stains on collars, sheets and pillow cases, bedspreads, towels, or wash cloths: These stains are often caused by the benzoil peroxide used in cosmetic products (including acne medicine). This chemical acts as a bleach, is very insoluble and hard to rinse off the body. It can permanently change colors of some dyes. The damage cannot be remedied, So it should be prevented.

When products containing this chemical must be used, white collars and household textiles may be a good choice. Stiff, coarse textures and/or dull colors in freshly laundered fabrics: Nonphosphate granular detergents can combine with hard water to leave behind a residue that can cause fabrics to become stiff and feel harsh. Avoid the problem by using a phosphate-based detergent, a heavy-duty liquid detergent or a nonprecipitating water conditioner with the nonphosphate granular detergent.

Soaking stiffened clothing in a solution of white vinegar and water (1 cup vinegar per gallon of water) may help restore them, however you should first test clothing for colorfastness to vinegar on a hidden seam allowance. Another way to restore this clothing is to treat as for yellow, gray, or general discoloration.

White powdery streaks on dark clothes: Powdery streaks on dark clothes are probably caused by undissolved detergent being incompletely rinsed out. Some nonphosphate detergents can deposit mineral hardness residue that shows as streaks. Avoid this problem by changing detergents or by adding detergent to the wash water first, then adding clothes and starting washer. Usually a repeat rinse and spin cycle with clear water will remove these streaks. White streaks on blue jeans: White streaks on blue jeans are probably not caused by un- dissolved detergent. Blue jeans are often dyed with indigo dye, which is a fugitive dye that bleeds in a water solution. As the washer spins, the edges where the fabric is folded get more abrasion and rougher treatment, causing the color to escape. Turning jeans wrong side out before laundering will reduce these white streaks and give more even fading. To avoid the natural fading that accompanies use of indigo, look for polyester/cotton jeans that are labeled colorfast. They will retain their dark blue.

Yellowing, graying, or general discoloration: This condition occurs when insufficient detergent issused for proper cleaning, wash water temperature is too low (especially for oil stains), too much detergent is used and insufficiently rinsed out, synthetics are washed with a light-duty detergent in cold water, or color is transferred from other non-colorfast items in the wash.

To refurbish clothing with this discoloration, wash in a permanent press cycle with hot wash water, a cool-down rinse, and a cup of water conditioner instead of detergent. If discoloration persists, repeat this procedure or wash again using the correct amount of detergent, an all- fabric bleach, or diluted liquid chlorine bleach if safe for fabric. The treatment of last resort for white items is treatment with a commercial color remover (Rit or Tintex). This reducing bleach must be used very carefully, as it will easily fade colors in any fabric it touches. If the yellow color is on silk, wool, or spandex it may be a result of fiber alteration due to improper use of chlorine bleach and is not removable.

SOAP AND DETERGENT

Water works with soap or detergent to get clothes clean. Soaps and detergents loosen dirt from clothes, keeping dirt in the wash water so it doesn't get back on the clothes. Most areas of New Mexico have hard water, so use detergent rather than soap. When soap mixes with hard water, they form a soapy scum that can turn clothes gray.

An all-purpose, heavy-duty detergent does a good job of cleaning most clothes. For delicate fabrics that require hand washing, use a gentler detergent such as dishwashing detergent. Most stores carry many brands of laundry products. All detergents contain similar types of cleaning ingredients. Detergent powders are the most common laundry product and are the least expensive. Liquid detergents are made especially for washing clothes in cold water, although they work in hot water, too.

Some people like lots of suds in the wash water. However, too many suds can cut down on washing action and are hard to rinse out. Suds may also damage the washing machine. High-sudsing detergents are not recommended for use in front-loading washing machines. Low-sudsing detergents can be used in either front- or top-loading machines.

- Directions on the package tell how much soap or detergent to use in an average-sized washing machine with normally soiled clothes.
- Measure soap or detergent carefully. Using too much wastes money and may make so many suds that clothes don't get clean. If not enough detergent is used, body oil and dirt may not come out.

- It may be necessary to add more soap or detergent if using an extra-large machine, if the clothes are very dirty, or if the water is very hard.

White clothes and linens may turn gray if not enough soap or detergent is used or if clothes are not thoroughly rinsed. To get clothes white again, follow this procedure:

- Wash the clothes again in hot water.
- Add enough water softener to make the water feel slippery (about 1 cup).
- If the water becomes sudsy, the clothes were not rinsed enough. Wash them again adding only water softener.
- If the clothes do not whiten, add soap or detergent and rewash.

WATER

Good washing action requires plenty of room for clothes to move around, so never crowd the wash tub. The washing solution must be able to circulate through clothes to loosen and carry away soil. Modern fabrics need even more room to move around than other fabrics. Set the water-level control on the machine for the size of the load to be washed.

Hot Water

- Removes dirt from heavily soiled items.
- Kills more germs than cold water.
- Fades the dyes in some colored clothes.
- Tends to cause wrinkling in some modern fabrics such as permanent press.

Warm Water

- Usually gets lightly soiled clothes clean.
- Does not kill germs unless a disinfectant is added.
- Is safe for most colored clothes.

Cold Water

- Requires a cold water detergent to get clothes clean. If a cold water detergent is not available, dissolve detergent powder in hot water before adding it to the wash water.
- Requires more detergent than warm water to get clothes clean.
- Does not kill germs unless a disinfectant is added.
- Recommended for washing some delicate fabrics.

Water reaching the washing machine is cooler than in the water heater. For very hot water in the washer, it may be necessary to set the water heater temperature on high.

Water Types

There are two main types of water: hard and soft. When using soap rather than detergent, soft water cleans clothes better than hard water. Hard water, which contains mineral deposits, requires more soap or detergent, and clothes don't get as clean as they do in soft water. Minerals in hard water combine with soap to form a scum, which causes laundry problems. Detergents work well in soft or hard water unless the water is especially hard.

Water Softeners

Packaged powders can be purchased to soften water, or a water softening machine can be purchased or rented. Figure out how much the powders would cost and compare that with the price of a machine. Water softening machines use salt to remove minerals from the water, so the salt content of household water is increased by the softening process. Persons who are on low sodium diets may prefer to use packaged water softeners.

Packaged water softeners come in two types: precipitating and non-precipitating softeners (water conditioners). Precipitating water softeners settle out the minerals that make water hard.

The water becomes cloudy. These softeners work better in non-automatic wringer washers than in automatics because you lift the clothes out of the water instead of spinning the water out of the clothes.

When using a powder, follow directions on the package. Add the softener to the water before adding soap or detergent. Add enough softener to make the water feel slippery. For best results use a water softener in the rinse water, too. Mechanical water softeners soften water as it flows through the machine. Follow operating instructions carefully and replenish salt as recommended.

CLOTHING MAINTENANCE

Clothing, once manufactured, suffers assault both from within and from without. The human body inside sheds skin cells and body oils, and exudes sweat, urine, and feces. From the outside, sun damage, damp, abrasion, dirt, and other indignities afflict the garment. Fleas and lice take up residence in clothing seams.

Well-worn clothing, if not cleaned and refurbished, will smell, itch, look scruffy, and lose functionality (as when buttons fall off and zippers fail).In some cases, people simply wear an item of clothing until it falls apart.

Cleaning leather presents difficulties; one cannot wash bark cloth (tapa) without dissolving it. Owners may patch tears and rips, and brush off surface dirt, but old leather and bark clothing will always look old. But most clothing consists of cloth, and most cloth can be laundered and mended (patching, darning, but compare felt).

Laundry, ironing, storage

Humans have developed many specialized methods for laundering, ranging from the earliest "pound clothes against rocks in running stream" to the latest in electronic washing machines and dry cleaning (dissolving dirt in solvents other than water).

Many kinds of clothing are designed to be ironed before they are worn to remove wrinkles. Most modern formal and semi-formal clothing is in this category (for example, dress shirts and suits). Ironed clothes are believed to look clean, fresh, and neat. However, much contemporary casual clothing is made of knit materials that do not readily wrinkle and so do not have to be ironed. Some clothing is permanent press, meaning that it has been treated with a synthetic coating (such as polytetrafluoroethylene) that suppresses wrinkles and creates a smooth appearance without ironing.

Once clothes have been laundered and possibly ironed, they are usually hung up on clothes hangers or folded, to keep them fresh until they are worn. Clothes are folded to allow them to be stored compactly, to prevent creasing, to preserve creases or to present them in a more pleasing manner, for instance when they are put on sale in stores.

Many kinds of clothes are folded before they are put in suitcases as preparation for travel. Other clothes, such as suits, may be hung up in special garment bags, or rolled rather than folded. Many people use their clothing as packing material around fragile items that might otherwise break in transit.

Mending

In past times, mending was an art. A meticulous tailor or seamstress could mend rips with thread raveled from hems and seam edges so skillfully that the darn was practically invisible. When the raw material — cloth — was worth more than labor, it made sense to expend labor in saving it. Today clothing is considered a consumable item. Mass-manufactured clothing is less expensive than the time it would take to repair it. Many people prefer to buy a new piece of clothing rather than to spend their time mending old clothes. But the thrifty still replace zippers and buttons and sew up ripped hems.

The Life Cycle of Clothing

Used, no-longer-wearable clothing was once desirable raw material for quilts, rag rugs, bandages, and many other household uses. It could also be recycled into paper. Now it is usually thrown away. Used but still wearable clothing can be sold at consignment shops, flea markets, online auction, or just donated to charity. Charities usually skim the best of the clothing to sell in their own thrift stores and sell the rest to merchants, who bale it up and ship it to poor Third World countries, where vendors bid for the bales and then make what profit they can selling used clothing.

Early 21st-century Clothing Styles

Western fashion has, to a certain extent, become international fashion, as Western media and styles penetrate all parts of the world. Very few parts of the world remain where people do not wear items of cheap, mass-produced Western clothing. Even people in poor countries can afford used clothing from richer Western countries.

However, people may wear ethnic or national dress on special occasions or if carrying out certain roles or occupations. For example, most Japanese women have adopted Western-style dress for daily wear, but will still wear silk kimonos on special occasions. Items of Western dress may also appear worn or accessorized in distinctive, non-Western ways. A Tongan man may combine a used T-shirt with a Tongan wrapped skirt, or tupenu. Western fashion, too, does not function monolithically. It comes in many varieties, from expensive haute couture to thrift store grunge.

Regional styles

- Clothing of Europe and Russia
- Clothing in the Americas
  — South American fashion
  — United States mainstream fashion

For example: "Catalogue" fashion, regional styles such as preppy or Western wear.

- United States alternative fashion

These fashions are often associated with fans of various musical styles.

See also Gothic fashion, Hippie, Grunge, Hip-hop, and Fetish-wear

- Clothing in Asia
- Clothing in Africa
- Clothing in Oceania

CLOTHING INDUSTRY

The clothing industry is concentrated outside of western Europe and America, and garment workers often have to labor under poor conditions. Coalitions of NGO's, designers (Katharine Hamnett, American Apparel, Veja, Edun...) and trade unions like the Clean clothes campaign (CCC) seek to improve these conditions as much as possible by sponsoring awareness-raising events, which draw the attention of both the media and the general public to the workers' conditions.

Clothes and Personality

Sometimes people can struggle to buy clothes because they feel uncertain of their wardrobe personality. They may feel as though they should express themselves in a certain way because it is fashionable, when in fact it makes them feel uncomfortable. Clarifying your wardrobe personality can help to give you a clearer focus and make you feel more confident about who you are. Style analysis divides wardrobe personalities into four categories but it is certainly possible to be a mixture of different personalities and to change over time.

Dramatic

Dramatics opt for clothes that get them noticed. They have a strong definite image and are not afraid of making bold statements. A typical dramatic will be drawn to bold colours and striking combinations, they may enjoy high fashion and looking funky. They embrace extremes and love to dress up one day and be completely casual the next but their main aim is to turn heads. Some dramatics enjoy designer labels but many enjoy expressing their individuality by shopping in second hand shops and car boot sales. For dramatics, shopping for clothes is like hunting, the pursuit of that unique item is thrilling. Such dramatic tastes follow through into jewellery, accessories, hairstyles and make-up.

For a dramatic, fading into the background is not an option but they need to be careful not to give the wrong impression at formal occasions. At these times, they may need to make sure their appearance is conveying the right message. They may also have a tendency to spend lavishly and have wardrobes upon wardrobes of clothes.

Classic

Classics tend to be much more quietly self confident than their dramatic counterparts. Their choice of clothing can be conservative to reflect their reserved personalities, although some classics can be highly dynamic and they utilize their classic demeanour to convey precision and strength. Classics have impeccable taste and aim for elegance rather than bold statements. They will often purchase extremely expensive clothing that will last for years. Their look is timeless. They opt for clean lines, well-cut clothes and blended colours. People will often comment that they look expensive. Their choice of jewellery is subtle and understated and so is their make-up and hair. Classics are immaculately groomed at all times and those around them eye them enviously in awe of their flawless appearance. For a classic, dignity is everything because it helps them to feel in control. The thought of wearing a tracksuit fills them with horror and they will even look immaculate doing the gardening. They may sometimes feel over dressed as casual is not in their vocabulary and shopping can be frustrating when classic styles are nowhere to be found.

Romantic

Luxurious romantics simply adore expressing their femininity. Romantics love to look attractive and rather than opt for clean lines and minimal detail like their classic counterparts, they are attracted like magpies to soft luxurious fabrics, beading and detailing. Aesthetics are everything to these beautiful creatures and they are attracted to anything that communicates luxury and expense. You can often smell a romantic before you can see them and only an expensive range of make-up will do.

They have an inexhaustible budget for grooming and will think nothing of booking in for a manicure, pedicure and a facial on a regular basis. For a luxurious romantic, indulgence is the key to happiness but they must be careful not to overspend or overdo the detail, perfume or make-up. Artistic romantics however are creative and sensitive and express their flair through choosing ornate detailing, beautiful fabrics and textures. Their appearance is more artistic than the luxurious romantic. You will often find an artistic romantic in a haberdashers or an ethnic shop.

Natural

For a Natural, comfortable, practical clothing is of utmost important. Their personality is easy going, down to earth, friendly and frank. Naturals are not materialistic; they see formal clothes as too restrictive and fussy accessories as pointless. A naturals wardrobe is minimal and practical and many live in jeans or tracksuit bottoms. They tend to be physical people and may lead very active lifestyles. Some naturals do not own a make-up bag or a jewellery box and their hair is in an easy to keep style.

Sometimes extreme naturals can lack the self-esteem to express themselves and often do not prioritise themselves as important enough to indulge and may be a natural through self-neglect. Extreme naturals need to be careful not to present themselves poorly in formal environments It is important to accept your wardrobe identity and express who you really are but don't be afraid to change and reflect different facets of your personality if that is what you desire to do.

Clothes Hanger

The Original SuperHanger design features make it an extremely effective garment hanger. The SuperHanger is the ultimate closet space saver allowing clothes to hang smoothly and effortlessly. Save time and money by consolidating coordinates and accessories, ensembles, complete outfits (including shoes, purses, and jewelry). Match up dresses with belts and hats for quick and easy access as to what to wear when pressed for time and in a hurry. Hanging garments on the SuperHanger will avoid time consuming chores such as ironing, steaming, unnecessary laundry, and trips to the dry cleaners. Garments retain their original shape, prevent wrinkles, and avoid slippage when properly hung on the SuperHanger. Sizable eye-catching location for branding allowing a clear image for repeated name recognition to build awareness and increase sales. Custom colors available including pearls, translucent, frosted colors, and simulated textures (i.e. marble, wood, granite, etc.). Logo area measuring 1 ¼" H x 3 ¾ W. Labels and embossing available.

A clothes hanger, or coat hanger, is a device in the shape of:

- Human shoulders designed to facilitate the hanging of a coat, jacket, sweater, shirt, blouse or dress in a manner that prevents wrinkles, with a lower bar for the hanging of trousers or skirts.
- Clamp for the hanging of trousers or skirts.

Both types can be combined in a single hanger.

There are two basic types of clothes hangers, the wire hanger, a simple loop of wire in a flattened triangle shape, with the wire continuing into a hook at the top, and the wooden hanger, a flat piece of wood cut into a boomerang-like shape, and with the edges sanded down to prevent damage to the clothing, with a hook, usually of metal, protruding from the point.

Some wooden hangers have a rounded bar from tip to tip, forming a flattened triangle, this bar is designed to hang the trousers belonging to the jacket. There are also plastic coat hangers, which mostly mimic the shape of either a wire or wooden hanger. Plastic coat hangers are also produced in smaller sizes to accommodate the shapes of children's clothes. There are also clothes hangers that have been padded with fine materials, such as satin, for expensive clothes, lingerie and fancy dresses. The soft, plush padding is intended to protect garments and keep them from getting shoulder dents from wire hangers.

Some hangers have clips along the bottom for suspending skirts. Dedicated skirt and trousers hangers may dispense with the triangular shape and just be a rod with clips. Specialized pant hanger racks may accommodate many pairs of trousers. A garment bag is designed to accommodate a clothes hanger with clothes such that the hanging hook protrudes from the top of the garment bag.

Some historians believe President Thomas Jefferson invented the wooden clothes hanger. However, today's most used hanger, the wire hanger, was inspired by a coat hook that was invented in 1869 by O. A. North of New Britain, Connecticut. An employee of the Timberlake Wire and Novelty Company, Albert J. Parkhouse has also been credited with the invention. In 1932, Schuyler C. Hulett patented an improved design, which used cardboard tubes mounted on the upper and lower parts of the wire to prevent wrinkles, and in 1935, Elmer D. Rogers added a tube on the lower bar, which is still used.

A wire clothes hanger was also a featured prop in a central scene in the 1981 movie Mommie Dearest, in which Joan Crawford, played by Faye Dunaway, went into the room of her daughter, Christina, at night while the girl was sleeping to admire the clothes hanging nicely in her closet. She then became enraged upon discovery that Christina had once again used a cheap wire hanger, instead of a nice one as she had instructed her to use, woke her up and gave her a thrashing. Joan's constant cry of "No wire hangers, ever!" quickly worked its way into pop culture and the phrase is now synonymous with those who tend to always want to be in control.

Clothes hangers are also now made of, though rarely, a rubber substance. Foldable clothes hangers that are designed to be inserted through the collar area to make it easier to use, and reduce stretching are a new, yet useful variation on traditional clothes hangers.

Unintended Uses

Wire clothers hangers are often used as cheap, easily available sources of semi-tough wire for all sorts of home projects. Many do it yourself, and children's projects use wire hangers as holders of various types, from keeping a brake caliper from dangling on its hose, to securing a gate on a bird cage. Once sanded wire hangers also find uses as conducting wire for uses as varied as hotwiring cars to games to test hand steadiness.

Wire clothes hangers are infamous as the symbol of illegal, and self-inflicted, abortion, for which uncoiled and straightened out wire hangers have been used as a tool. Actress Whoopi Goldberg appeared at a pro-choice rally brandishing a clothes hanger, which caused much controversy. They are also commonly used to forcibly gain entry into automobiles; however with the increased complication of car locks and entry systems, such methods are often unsuccessful on newer cars.

CLOTHES DRYER

A clothes dryer or tumble dryer is a major household appliance that is used to remove the residual moisture from a load of clothing and other textiles, generally shortly after they are cleaned in a washing machine. Most dryers consist of a rotating drum called a tumbler through which heat is circulated to evaporate the moisture from the load. The tumbler is rotated relatively slowly in order to maintain space between the articles in the load. In most cases, the tumbler is belt-driven by an induction motor.

Fuel Types

There are generally two classes of rotating dryers: electric and gas. Both of these refer to the method used to raise the temperature of the air flowing through the tumbler.

The electric dryer generally uses a coiled wire that is heated with electric current. The amount of electric current is varied to adjust the air temperature. In the United States, electric dryers typically have a 4-wire NEMA 14-30 plug, rather than the 3-wire NEMA 5-15 plug used by most appliances, and need a 30-ampere, 240-volt centertapped single-phase circuit. The gas dryer employs a gas burner that burns natural gas, propane, or butane to form a jet of hot gases that are directed into a venturi chamber, which uses Bernoulli's principle to pull in ambient air and raise its temperature. The air temperature can be altered by adjusting the size of the gas flame or, more commonly, by merely extinguishing it and relighting it.

Traditional Dryers

Traditional dryers continuously draw in the cool, dry, ambient air around them and heat it before passing it through the tumbler. Afterwards, the resulting hot, humid air is simply vented outside to make room for more dry air to continue the drying process. The traditional design makes no effort to recycle the heat put into the load, and so is quite inefficient. Nevertheless, the basic design is simple, reliable, and cheap.

Ventless Dryers

Spin Dryers

These machines simply spin their drums faster than a typical washer could in order to extract additional water from the load. They may remove more water in two minutes than a heated tumbler dryer can in twenty, thus saving significant amounts of time and energy. Although spinning alone will not completely dry clothing, this additional step saves a worthwhile amount of time and energy for large laundry operations such as those of hospitals or hotels.

Condensation Dryers

Just as in a normal dryer, condensation dryers pass heated air through the load. However, instead of exhausting this air, the dryer uses a heat exchanger to cool the air and condense the water vapor into either a drain pipe or a collection tank. Afterwards, this air is run through the loop again. The heat exchanger uses ambient air as its coolant, therefore the heat produced by the dryer will go into the immediate surroundings instead of the outside.

Because the heat exchange process simply cools the internal air using ambient air, it will not dry the air in the internal loop to as low a level of humidity as the fresh, ambient air. As a consequence of the increased humidity of the air used to dry the load as well as the increased complexity of the design, this type of dryer requires more time and energy than a traditional dryer. However, it is a valid option where long, intricate ducting would be required to vent a traditional dryer.

Heat Pump Dryers

Whereas condensation dryers use a passive heat exchanger cooled by ambient air, these dryers use a heat pump. The hot, humid air from the tumbler is passed through a heat pump where the cold side condenses the water vapor into either a drain pipe or a collection tank and the hot side reheats the air. In this way not only does the dryer avoid the need for ducting, but it also conserves much of its heat within the dryer instead of exhausting it into the surroundings. Heat pump dryers can therefore use less than half the energy required by either condensation or traditional dryers.

As with condensation dryers, the heat exchanger will not dry the internal air to as low a level of humidity as the ambient air. With respect to ambient air, the higher humidity of the air used to dry the clothes has the effect of increasing drying times; however, because heat pump dryers conserve much of the heat of the air they use, the already-hot air can be cycled more quickly, possibly leading to shorter drying times than traditional dryers, depending on the model.

Mechanical Steam Compression Dryers

A new type of dryer in development, these machines are a more advanced version of heat pump dryers. Instead of using hot air to dry the clothing, mechanical steam compression dryers use water recovered from the clothing in the form of steam. First, the tumbler and its contents are heated to 100 °C. The wet steam that results purges the system of air and is the only remaining atmosphere in the tumbler.

As wet steam exits the tumbler, it is mechanically compressed (hence the name) to extract water vapor and transfer the heat of vaporization to the remaining gaseous steam. This pressurized, gaseous steam is then allowed to expand and is superheated before being injected back into the tumbler where its heat causes more water to vaporize from the clothing, creating more wet steam and restarting the cycle.

Like heat pump dyers, mechanical steam compression dryers recycle much of the heat used to dry the clothes, and they operate in a very similar range of efficiency as heat pump dryers. Both types can be over twice as efficient as traditional dryers. However, the considerably higher temperatures used in mechanical steam compression dryers result in drying times on the order of half as long as those of heat pump dryers.

Static

Clothes dryers can cause static cling, through the triboelectric effect. Contrary to Internet rumors, the first tumble dryer was not invented by American George T. Sampson. A hand-cranked version was created in 1799 by a Frenchman named Pochon. Sampson's United States patent (number 476,416), which he received on June 7, 1892, was for an improved rack for holding wet clothes near a heat source. Electric tumble dryers appeared in the 20th century.

Ecology

Energy Star, a United States government program concerned with energy efficiency in consumer products, does not rate clothes dryers, because most of the clothes dryers marketed in the U.S. consume similarly massive amounts of energy.

CLEANING WITH MICROFIBRE CLOTH

Two publications have reported positively on the use of microfibre technology. These are *The Impact of Microfibre Technology on the Cleaning of Health care Facilities* (Association of Health care Cleaning Professionals, revised 2006) *and An Integrated Approach to Hospital Cleaning: Microfibre Cloth and Steam Cleaning Technology* (Department of Health, May 2007).

It is recommended that all Trusts should read these reports and consider implementing the routine use of microfibre technology.

Microfibre mops and cloths are made of composite synthetic fibres which are extremely fine, and which are engineered to have a large surface area. This gives a much greater effective cleaning capacity, and enables the extremely efficient removal of microscopic particles.

The small size of the microfibres enables them to reach into microscopic crevices in surfaces. Additionally, the microfibres are naturally statically charged. This combination of electrostatic attraction and capillary action allows the removal of a far greater number of contaminant particles than conventional mops and cloths.

Many hospitals using microfibre cloths have reported that the time taken to perform cleaning tasks has reduced, and that the introduction of microfibre has been followed by improvements in measured cleaning scores. There are limitations on the use of microfibre cloths. They are designed to be used dampened only with water, and therefore should not be used in conjunction with chlorine-based disinfectant cleaners or other biocides. Used cloths will contain potentially harmful contaminants.

A separate cloth should be used for each patient area and cloths must be thermally disinfected. Microfibre cloths are less effective when used on old and damaged surfaces because of repeated snagging and perform best in the routine maintenance of surfaces which are not heavily soiled.

There are now many commercially available microfibre products. Not all of these are of the highest grade and careful consideration should be given to the relative merits of products on the market. Once a product is chosen, a detailed implementation plan should be created, including thorough training for supervisors and

staff with cleaning responsibilities. The method statement provided here is provided as an example. It should be noted that different manufacturers may make different recommendations for how their products should be used. Organisations should ensure that practice is consistent with the manufacturer's recommendations and the advice of the local Infection Control Team. Microfibre cloths are designed to be used as part of a well coordinated and tightly controlled cleaning system. Although some hospitals do currently use microfibre cloths for one task in isolation, such as high-dusting, this may not produce an optimum result. For this reason, the method statement given below is for the integrated use of microfibre for a range of tasks in an area.

CLEANING WITH MICROFIBRE CLOTHS

Equipment and materials required:

- Colour-coded domestic gloves;
- Cleaning trolley designed for use with microfibre cleaning system;
- Dust-control mop and handle;
- Damp-mopping mop and handle;
- Colour-coded labelled net bag containing clean microfibre flat mops;
- Colour-coded labelled net bag containing clean microfibre cloths;
- High-dusting tool, with telescopic attachment if required;
- Microfibre sleeve for high-dusting tool;
- Colour-coded dustpan and brush;
- Laundry bag for used microfibre cloths;
- Laundry bag for used microfibre flat mops;
- Labelled spray gun containing general purpose detergent or other compatible cleaning product;
- Labelled spray gun containing cold water;
- Warning signs.

Method:

- Wash hands and put on gloves.
- Plan work route and temporarily remove potential obstacles to a new, safe location. The area to be cleaned at one time should be no larger than half of a six-bedded patient bay.
- Display warning signs.
- Prepare the cleaning solution in the bucket, in strict accordance with the manufacturer's instructions and with your training. Do not mix chemicals and only use a cleaning product provided by your employer.
- Assemble items to be used on the cleaning trolley: place clean microfibre cloths in the appropriate container. Place clean microfibre flat mops in the mop container. Place net bags over the laundry bags ready to receive used cloths and mops. Impregnate mops with the correct amount of water, following the manufacturer's instruction.
- Begin by high-dusting the area, using the high-dusting tool and microfibre sleeve (refer to high-dusting method statement for general guidance on high-dusting).
- On completion of high-dusting, remove microfibre sleeve and place in cloths laundry bag.

- Damp-dust all surfaces. Spray a small amount of cold water onto the surface to be cleaned and wipe with a single firm stroke. Turn and refold the cloth each time the surface of the cloth being used becomes full of dust. As a guide, a standard sized cloth can be turned and folded so as to give eight effective cleaning surfaces. One cloth should normally be sufficient to clean each area.
- Work systematically from higher areas to low, taking care to damp-dust the edges and undersides of all surfaces after the tops. Where extendable items, such as bedside tables, are to be damp-dusted, extend them before beginning to work.
- Dust-control the area. Attach a microfibre dust-control cloth to the dust-control tool head.
- Manually pick up any larger items such as sweet wrappers and tissues.
- Starting at an edge, dust the area using an overlapping figure-of-eight pattern, taking care to go right up to every edge. The dust-control tool should remain in contact with the floor at all times.
- When you have finished, carefully remove cloth and fold the cloth so that dust is not dispersed. Use the dustpan and brush to collect any dust and debris which has collected at your finishing point.
- Damp mop the area. Attach a microfibre flat mop head to the mopping tool.
- Starting at an edge, mop the area using an overlapping figure-of-eight pattern, taking care to go right up to every edge. The mop head should remain in contact with the floor at all times.
- Greasy or stubborn deposits may require use of the cleaning solution. Spray a small amount of solution onto the soiled area and leave in contact for a short time, then repeat damp-mopping as above.
- Return any items moved, to the clean surface when it is dry.
- Move onto the next work area and repeat points 1-17 above.

19. On completion, clean and dry all equipment and store safely and tidily in a secure storage area, segregated according to colour-coding where appropriate.

- Remove gloves and wash hands.

Ward pantry cleaning:

- Ice-making machine
- Washing up
- Mechanical waste disposal unit
- Microwave
- Oven cleaning
- Refrigerator
- Kitchen sink
- Regeneration unit – routine daily clean

ICE-MAKING MACHINE

Equipment and materials required:

- Colour-coded bucket;
- Colour-coded cloths;
- Colour-coded domestic gloves;
- Mon-abrasive pad;
- Warning signs;
- General purpose detergent.

Additional Notes:

- Notify the nurse in charge that you are intending to carry out this procedure and delay task to allow ice to be decanted into a suitable receptacle, if it is required urgently.
- Take care to mop up any spillages of water immediately, should they occur while you are carrying out this task.

Method:

- Wash hands and put on gloves.
- Display the warning signs.
- Prepare the cleaning solution in the bucket, in strict accordance with the manufacturer's instructions and with your training. Do not mix chemicals and only use a cleaning product provided by your employer.
- Disconnect the ice-making machine from the mains socket.
- Empty the ice bin and dispose of the ice.
- Following the manufacturer's instructions, carefully remove all detachable parts.
- Dampen or rinse the cloth in the cleaning solution and wring so that it is fairly dry.
- Clean the detachable parts and all the surfaces of the machine, paying particular attention to the dispensing, overflow and draining areas.
- Rinse cloth regularly and change the cleaning solution when it becomes soiled.
- Empty and clean the bucket and refill with warm water.
- Using a new cloth, rinse the detachable parts and all surfaces.
- Allow to dry, and carefully reassemble the machine, following the manufacturer's instructions.
- On completion, clean and dry all equipment and store safely and tidily in a secure storage area, segregated according to colour-coding where appropriate.
- Remove gloves and wash hands.
- With dry hands, plug the machine into the mains socket and test for correct operation. If it does not operate correctly, report this to your supervisor.

WASHING UP

Note: Because very high temperature water should be used for the cleaning of shared use patient crockery and cutlery, it is strongly advised that a mechanical dishwasher is used for this process. Where this is impossible, the task should be risk-assessed and a local policy developed.

Equipment and materials required:

- Colour-coded cloth;
- Colour-coded domestic gloves;
- Colour-coded polypropylene scrubbing brush;
- Abrasive pad;
- General purpose detergent;
- Warning signs.

Method:

- Wash hands and put on gloves.
- Display warning signs.
- Remove all residual waste from crockery and place in the mechanical waste disposal unit, or in a dedicated waste food bag ready for disposal.
- Sort the most heavily soiled items from the rest and pre-wash them under a running tap.
- Fill the sink to two-thirds full with hot water, then add general purpose detergent, in accordance with the manufacturer's instructions and with your training.
- Immerse soiled items in the hot cleaning solution and clean off visible soiling. Greasy or stubborn deposits may require removal with the abrasive pad.
- Place crockery and cutlery in the dishwasher tray, put tray into dishwasher and start dishwashing cycle.
- Dispose of the cloth, clean and dry all equipment and store safely and tidily in a secure storage area, segregated according to colour-coding where appropriate.
- Remove gloves and wash hands.
- When cycle is finished, wash hands and put on clean gloves and neatly store crockery and cutlery for next use.

MECHANICAL WASTE DISPOSAL UNIT

Equipment and materials required:

- Colour-coded bucket;
- Colour-coded cloths;
- Colour-coded domestic gloves;
- Non-abrasive pad;
- Warning signs;
- General purpose detergent or degreaser.

Method:

- Wash hands and put on gloves.
- Display the warning signs.
- Turn on the waste disposal unit for twenty seconds to dispose of any residual waste.
- Turn off the electricity supply at the mains socket.
- Following the manufacturer's instructions, carefully remove all detachable parts.
- Prepare the cleaning solution in the mop bucket, in strict accordance with the manufacturer's instructions and with your training. Do not mix chemicals and only use a cleaning product provided by your employer.
- Dampen or rinse the cloth in the cleaning solution and wring so that it is fairly dry.
- Carefully clean the detachable parts and all the surfaces of the machine. For greasy deposits the non-abrasive pad may be required.
- Clean the surrounding areas.
- Rinse cloth regularly and change the cleaning solution when it becomes soiled.
- Empty and clean the bucket and refill with warm water.
- Using a new cloth, rinse the detachable parts and all surfaces.
- Allow to dry and carefully replace the detachable parts, following the manufacturer's instructions.
- On completion, clean and dry all equipment and store safely and tidily in a secure storage area, segregated according to colour-coding where appropriate.
- Remove gloves and wash hands.
- With dry hands, plug the machine into the mains socket and test for correct operation. If it does not operate correctly, report this to your supervisor.

MICROWAVE

Equipment and materials required:

- Colour-coded bucket;
- Colour-coded cloths;
- Colour-coded domestic gloves;
- Non-abrasive pad;
- Warning signs;
- General purpose detergent.

Method:

- Wash hands and put on gloves.
- Display the warning signs.

- Prepare the cleaning solution in the bucket, in strict accordance with the manufacturer's instructions and with your training. Do not mix chemicals and only use a cleaning product provided by your employer.
- Disconnect the microwave from the mains socket.
- Dampen or rinse the cloth in the cleaning solution and wring so that it is fairly dry.
- Clean the microwave, working from outside to inside, paying particular attention to the "ceiling" of the inside and the inside of the door. For greasy or hardened deposits, the non-abrasive pad may be required.
- Rinse cloth regularly and change the cleaning solution when it becomes soiled.
- Empty and clean the bucket and refill with warm water.
- Using a new cloth, rinse all surfaces with a fairly dry cloth.
- Allow to dry with the door open.
- On completion, clean and dry all equipment and store safely and tidily in a secure storage area, segregated according to colour-coding where appropriate.
- Remove gloves and wash hands.
- With dry hands, plug the machine into the mains socket and test for correct operation. If it does not operate correctly, report this to your supervisor.

OVEN CLEANING

Equipment and materials required:

- Colour-coded bucket;
- Colour-coded cloths;
- Colour-coded domestic gloves suitable for chemical resistance and complying with the PPE Directive (89/686/EEC);
- Colour-coded dustpan and brush;
- Gauntlet gloves;
- Colour-coded scraper;
- Colour-coded labelled spray bottle;
- Eye goggles;
- Non-abrasive pad;
- Paper towel roll;
- Warning signs;
- General purpose detergent and caustic cleaner if heavy soiling present;
- Spray on oven cleaner chemical product.

Method:

- Wash hands and put on gloves.
- Display the warning signs.

- Paper towels should be placed along the join between the floor and oven to absorb leaking cleaning fluid.
- Take care to check the temperature of the oven carefully before attempting to clean it.
- Prepare the general purpose detergent cleaning solution in the bucket, in strict accordance with the manufacturer's instructions and with your training. Do not mix chemicals and only use a cleaning product provided by your employer.
- Disconnect the oven from the mains socket and ensure that the oven controls are set to "off".
- Carefully test the temperature of the oven by placing your hand into the oven. Delay starting the task if the oven is too hot to clean safely.
- Remove the shelves to a safe cleaning area.
- Carefully remove any loose, flaky soiling with the scraper. Hold the scraper parallel to the oven surface and gently push soiling loose. Collect from the floor of the oven with the dustpan and brush.
- Dampen or rinse the cloth in the cleaning solution and wring so that it is fairly dry.
- Clean the oven, working from outside to inside, paying particular attention to the "ceiling" of the inside, the inside of the door and the joint between the door and the oven. For greasy or hardened deposits the non-abrasive pad may be required.
- Rinse cloth regularly and change the cleaning solution when it becomes soiled.
- Empty and clean the bucket and refill with warm water.
- Using a new cloth, rinse all surfaces with a fairly dry cloth.
- Allow to dry.
- Move to the shelves and repeat points 7 to 13.
- Where there is heavy soiling which resists cleaning with general purpose detergent, a caustic oven cleaning chemical should now be used. Remove gloves and wash hands.
- Put on gauntlet gloves and eye goggles and carefully spray small amounts of the product onto the soiled areas. Leave in contact in strict accordance with manufacturer's instructions. Use the non-abrasive pad to clean off the soiling.
- Repeat points 11) to 14), taking care to ensure that the chemical residue is removed. Change water and cloth when they become dirty.
- Replace the shelves in the oven.
- On completion, clean and dry all equipment and store safely and tidily in a secure storage area, segregated according to colour-coding where appropriate.
- Remove gloves and wash hands.

- With dry hands, turn the oven on and test for correct operation. If it does not operate correctly, report this to your supervisor.

CLEANING WITH PRESSURISED STEAM – ROUTINE CLEANING

Equipment and materials required:

- Colour-coded, heat resistant gauntlet gloves suitable for chemical resistance and complying with the PPE Directive (89/686/EEC);
- Eye goggles;
- Colour-coded cloths;
- Pressurised steam cleaner;
- Steam cleaner accessories;
- Warning signs.

Method:

- Plan work route and when necessary, temporarily move items that may obstruct you to a new, safe location. Identify a suitable drain for disposal of the dirty water.
- Wash hands and put on heat resistant gauntlet gloves and eye goggles.
- Display warning signs.
- Fill the steam cleaner with clean water, following the manufacturer's instructions and your training.
- Attach the general purpose nozzle head.
- Unwind the cable, plug into a mains socket and turn on the machine.
- Wait for the water to reach the required temperature, following the manufacturer's instructions and your training. This will normally take between 4 and 8 minutes depending on the machine used.
- When the machine is ready to use, begin cleaning. Starting with the highest areas and moving to the lowest, clean ledges and surfaces in a 1-2 metre section, taking care not to overstretch. Use the suction cleaning function following the manufacturer's instructions and your training.
- Repeat the process, moving systematically around the room section by section.
- When completed, turn off the machine, fit the crevice cleaning attachment, or the smallest directional nozzle and turn the machine back on. Clean the joint between the wall and the floor and other awkward areas, which have resisted the general steam cleaning, moving systematically around the room as before.
- When completed, turn off the machine, fit the floor cleaning attachment and turn the machine back on. Clean the main surface of the floor, beginning at the point furthest from the door and progressing towards it.

- With dry hands, remove the plug from the mains socket and rewind the electricity cable.
- Empty the dirty water tank in identified drain.
- Collect all accessories.
- Clean the steam cleaner and accessories.
- Store the cleaned equipment safely and tidily in the secure storage area, segregated according to colour-coding where appropriate.
- Return any items moved to their original positions.
- Remove gloves and wash hands.

CLEANING WITH PRESSURISED STEAM – DEEP CLEAN OF EQUIPMENT

Note: This method statement is for the detailed cleaning of multiple pieces of moveable equipment, such as commodes, wheeled trolleys and drip stands, carried out in a suitable cleaning area. This will normally be done as part of planned periodic deep cleaning. To maximise efficiency, the operative carrying out the steam cleaning should remain in place while colleagues transport items to be cleaned to and from their normal locations. Care should be taken to segregate clean and dirty items. The same basic method is suitable for cleaning of single items of equipment in their normal locations.

Equipment and materials required:

- Colour-coded, heat resistant gauntlet gloves suitable for chemical resistance and complying with the PPE Directive (89/686/EEC);
- Eye goggles;
- Colour-coded cloths;
- Pressurised steam cleaner;
- Steam cleaner accessories;
- Warning signs.

Method:

- Identify a suitable work area, close to a drain for disposal of the dirty water. Designate and sign areas for items awaiting cleaning and cleaned items, and ensure that these are kept separate. The area for cleaned items should be well away from the area where cleaning is to be performed.
- Wash hands and put on heat resistant gauntlet gloves and eye goggles.
- Display warning signs.
- Fill the steam cleaner with clean water, following the manufacturer's instructions and your training.
- Attach the general purpose nozzle head.
- Unwind the cable, plug into a mains socket and turn on the machine.
- Wait for the water to reach the required temperature, following the manufacturer's instructions and your training. This will normally take

between 4 and 8 minutes depending on the machine used.

- When the machine is ready to use, begin cleaning. Clean the main surfaces, undersides and edges of the item of equipment.
- When completed, turn off the machine, fit the crevice cleaning attachment, or the smallest directional nozzle, and turn the machine back on. Carefully clean any crevices or awkward joints, using the suction function. Repeated cleans may be necessary where there is heavy soiling. Repeat the process for the wheels and for any other areas which have resisted general steam cleaning.
- Move the item of equipment to the identified area for cleaned items.
- Move onto the next item of equipment, and repeat points 5 to 10.
- When all items are completed, with dry hands remove the plug from the mains socket and rewind the electricity cable.
- Empty the dirty water tank in identified drain.
- Collect all accessories.
- Clean the steam cleaner and accessories.
- Store the cleaned equipment safely and tidily in the secure storage area, segregated according to colour-coding where appropriate.
- Remove gloves and wash hands.

BARRIER-NURSED ROOMS

Note: The use of combined hypochlorite disinfectant cleaning products, which provide a solution of 1.000 (one thousand) parts per million available chlorine combined with a detergent property, has largely replaced the previous use of general purpose detergent solution followed by an application of a disinfectant product. The method statement reflects this.

Barrier-nursed rooms will have a clear sign, prominently displayed, stating which items of protective disposable wear are required by all those entering the room. Cleaning staff should report daily to the nurse in charge to receive any new instructions

Equipment and materials required:

- Colour-coded bucket;
- Colour-coded disposable cloths;
- Colour-coded domestic gloves suitable for chemical resistance and complying with the PPE Directive (89/686/EEC);
- Single-use gloves;
- Disposable apron;
- Colour-coded mop handle;
- Colour-coded mop bucket;
- Colour-coded mop head;
- Colour-coded high-duster;
- Dust-control tool;

- Dust-control head;
- Cleaning trolley;
- Warning signs;
- Disinfectant cleaning product giving sodium hypochlorite solution of strength 1,000 parts per million available chlorine;
- Product dilution container.

Method:

- Wash hands and put on gloves.
- Prepare the cleaning solution. The ventilation of the area in which you are working must be thorough; if there is no window, the door should be left open. Make up the solution in the product diluting container, never directly in the bucket. Use cold water only. When prepared, decant carefully into the buckets.
- Wash hands, put on single-use gloves and other protective wear required (see Health and Safety notes) and enter room.
- Display the warning signs.
- High-dust the room (refer to high-dusting method statement).
- Dust-control the floor (refer to the dust-control method statement).
- Using a disposable cloth dampened in the disinfectant cleaning solution, begin cleaning surfaces, changing cloths and solution as they become soiled. Work in the following order: curtain tracks and high ledges; furniture and fittings; patient equipment, such as drip stands; alcohol gel, soap and paper towel dispensers (refer to replenishment method statements); bed frame; wash hand basin, shower and toilet (refer to wash room cleaning and sanitary cleaning method statements).
- Attach the mop head to the mop handle.
- Submerge the mop into the cleaning solution and remove excess using the wringer so that the mop is fairly dry.
- Mop the floor in 1-2 metre square sections.
- Mop edges with straight strokes and use a figure-of-eight pattern on the remainder of the section, turning the mop frequently. The floor should be fairly dry on completion.
- Move to the next section and repeat the process.
- Replace the mop head and solution as required throughout the cleaning process.
- On completion, remove the final mop head and place in the laundry bag, clean and dry all equipment and store safely and tidily in a secure storage area, away from other equipment; segregated according to colour-coding where appropriate.
- Remove and dispose of single-use gloves, apron and other protective wear. Wash hands.

5

Laundry and Dry Cleaning Management

LAUNDRY MANAGEMENT

Laundry work involves taking items of clothing or linen from wards and other patient areas, providing an appropriate cleaning service and returning items appropriately finished. Laundry operatives will operate steam cabinets, use steam and vacuum presses or finish items by hand. They sort for collection and pack items as required. They may also offer a repair service and operate a sewing machine. They will handle items in bulk.

Laundry can be:

- Items of clothing and other textiles that require washing,
- The act of washing clothing and textiles,
- The room of a house in which this is done.

History of laundry

Laundry was probably first done by immersing cloth in streams and letting the stream carry away the materials causing stains and smells. Laundry may still be done this way in some less industrialized areas and rural regions. Agitation helps remove the dirt, so the laundry is often rubbed, twisted, or slapped against flat rocks.

Various chemicals may be used increase the solvent power of water, such as the compounds in soaproot or yucca-root used by Native American tribes. Soap, a compound made from lye (from wood-ash) and fat, is an ancient and very common laundry aid. However, modern washing machines typically use powdered or liquid laundry detergent in place of soap.

When no streams were available, laundry was done in water-tight vats or vessels. Sometimes large metal cauldrons were filled with fresh water and heated over a fire; boiling water was even more effective than cold in removing dirt. The washboard, a corrugated slab of a hard material such as metal, replaced rocks as a surface for loosening soil. Once clean, the clothes were wrung out — twisted to remove most of the water. Then they were hung up on poles or clotheslines to dry, or sometimes just spread out on clean grass.

The Industrial Revolution

The Industrial Revolution completely transformed laundry technology. First came the mangle (wringer US), in the 18th century — two long rollers in a frame and a crank to revolve them. A laundry-worker took sopping wet clothing and cranked it through the mangle, compressing the cloth and expelling the excess water. The mangle was much quicker than hand twisting.

In the early 20th century inventors further mechanized the laundry process with various washing machines. Typically, these machines used an electrically powered agitator to replace tedious hand rubbing against a washboard. The earliest machines were simply a tub on legs, with a hand-operated mangle on top. Later the mangle too was electrically powered, then replaced by a perforated double tub, which spun out the excess water in a spin cycle. Laundry drying was also mechanized, with clothes dryers. Dryers were also spinning perforated tubs, but they blew heated air rather than water.

Modern Laundry

Washing machines and dryers are now fixtures in homes around the world. Apartment buildings and dormitories often have laundry rooms, where residents share washing machines and dryers. Usually the machines are set to run only when coins in appropriate amounts are inserted in a coin-slot. Those without home machines or access to laundry rooms must either wash their clothes by hand or visit a commercial laundromat.

LAUNDRY DETERGENT

Laundry detergent, or Washing Powder, is a substance which is a type of detergent that is added when one is washing laundry to help get the laundry cleaner. It is often colloquially called laundry soap or simply detergent or soap and it helps wash the fabric in a manner rather analogous to the way soap helps wash hands, other parts of the body, or other things cleaner than washing with water alone.

Laundry detergent has traditionally been a powdered or granular solid, but the use of liquid laundry detergents has gradually increased over the years, and these days use of liquid detergent equals or even exceeds use of solid detergent. Some brands also manufacture laundry soap in tablets and dissolvable packets, so as to eliminate the need to measure soap for each load of laundry.

Use of Laundry Detergent

Laundry detergent is used by almost every household in the developed world, so its production and sale has become an important industry. One can find many different brands of laundry detergent in the shops, and each one may claim to have any number of special qualities. Each brand has its own instructions on how it should be used and what amount to use written on the container it comes in. Some brands of laundry detergent purport to be more concentrated and can be added in smaller amounts. The detergent can be added onto the laundry or to the wash water at the start of the wash, or it can be added beforehand or soon after starting the wash into a special compartment in a washing machine made for that purpose, to be flushed into the wash by the wash water. Often bleach is a separate additive to the wash, but there are laundry detergents which have bleach already blended in with them.

All laundry detergents are soluble in water so as to make the water more effective at cleaning the laundry. The detergent does its work and is needed during the initial "Wash" cycle to separate the dirt or soil out from the fabric. The purpose of the rinses which follow is to rinse the detergent residue from the laundry as well as to remove the dirt suspended in the wash water by replacing the initial wash water with fresh water.

Separate stain pre-treatment products are sometimes available to be added directly to stained areas of fabrics. However, undiluted liquid laundry detergent or a paste of solid detergent made by mixing in a little water may be used to pre-treat heavily soiled or stained areas of the fabric by scrubbing in the detergent prior to the wash. Removal of certain types of, special oxidizer washing products have become available in liquid or solid granular form.

Such solid products often contain sodium percarbonate or sodium perborate and the liquid products often contain hydrogen peroxide. They are often marketed as containing "active oxygen". Some laundry detergent is specially sold for delicate woolen articles. Washing detergenty was invented by james Melvin Lawerence in 1994.

Containers and Sizes

Solid laundry detergent is commonly sold in cardboard boxes and plastic tubs. In many parts of the world, laundry detergent is also sold in single-use packets. The size of the boxes can vary from small single-use boxes sold from vending machines in laundromats to large economy-size boxes. In some cases, plastic measuring scoops have been included inside the boxes.

Liquid detergent is sold in plastic bottles, usually high density polyethylene or sometimes PET or other kinds. Again, various sizes are available. On large size bottles, a handle to carry the bottle is often pre-formed as part of the bottle. The bottle caps are often made large enough so they can be used as cups for measuring out the liquid detergent.

Contents

Detergent molecules are long and asymmetrical: one end attracts dirt while the other end attracts water, which is why detergent lifts dirt from wet clothes. Detergent molecules come in two forms, straight and branched - a difference that affects sudsing but not cleansing ability. The suds of the straight molecules of detergent break down quickly, while the suds of branched molecules break down slowly if at all.

A key ingredient in both solid and liquid laundry detergents is a surfactant. A surfactant is a substance which, when added to water, significantly reduces the surface tension of the water. This effect allows water to wash surfaces better. There are many different types of organic compounds which can function as surfactants. Most surfactants are thick, viscous liquids, but some are soft, waxy or greasy solids.

Surfactants typically have somewhat longer molecules which may or may not have an electric charge. Surfactants with uncharged molecules are non-ionic surfactants. Surfactants with positively charged molecules (or ions) are cationic surfactants. Surfactants with negatively charged molecules (or ions) are anionic surfactants. Surfactants with both positively and negatively charged part in the same molecule are zwitterionic surfactants. Most brands of laundry detergent have anionic or nonionic surfactants or a mixture of the two, although cationic surfactants have been used in laundry detergents. The use of cationic and anionic surfactants together is incompatible in the same detergent.

In powdered or granular solid detergents, the surfactant is soaked into the solid ingredients. In liquid laundry detergents, liquid or even solid surfactant are blended into the liquid detergent. There is usually a limit on how much liquid surfactant can soak into powder or granular solids before making the solid detergent mushy.

More liquid surfactant can usually be blended into a liquid detergent. The liquid detergents commonly contain at least some water to help liquefy the other additives and still have the detergent pour able. The liquid detergents may also have other solvent liquids, such as alcohol or a hydrotrope, to help blend all the additives together.

Laundry detergents may have ingredients to help control the pH of the wash water. For example, solid detergents usually contain sodium carbonate (soda ash) or sodium bicarbonate to maintain pH by neutralizing any acidic materials that may enter the wash water. Some other ingredients which solid detergents may have include sodium silicate or some form of sodium phosphate such as trisodium orthophosphate, monosodium orthophosphate, or a form of tripolyphosphate. In some locations, phosphate is no longer used as an additive due to environmental concerns.

Some laundry detergents have enzymes to help in stain removal, often enzymes produced by the bacteria Bacillus subtilis and Bacillus licheniformis. Some laundry detergents have fabric softeners. Perfume or color ingredients are sometimes added for better smell or to give a detergent some color.

Other brands, however, are left without these additives, marketed to those who avoid these because of allergies or individual preference. There are also detergents made with vegetable-based surfactants which are popular in health food stores. In the 1960s, detergent manufacturers waged an advertising battle over who had the longest lasting suds, and detergent compounds quickly appeared in the waterways. Suds began to appear in streams, rivers, lakes, and at the foot of Niagara Falls, piles of discolored detergent foam rose 8 feet high.

Detergents also containe phosphate additives to soften the water and thereby improve the effectiveness of the detergent molecules. It was noted that between 1940 and 1970 the amount of phosphates in city wastewater increased from 20,000 to 150,000 tons per year. With the increase in phosphates, algal blooms grew splendidly on the excess phosphorous and consumed the majority of all oxygen in the waters, killing fishes and plants.

PROCESS OF DRY CLEANING

Dry cleaning uses fluids to remove soil and stains from fabric. In fact, the term "dry cleaning" is misleading; it is called dry cleaning because the fluid contains little or no water and does not penetrate the fibers as water does. Among the advantages of dry cleaning is its ability to dissolve greases and oils in a way that water cannot. Drycleaning helps to return garments to a "like-new" condition using precautions to prevent shrinkage, loss of colour, and fabric distortion. The dry cleaning process begins with the pretreatment of spots and stains using special cleaning agents.

The garments are then loaded into a machine resembling an oversized front-loading home washer. Throughout the cleaning process the fluid is filtered or distilled to ensure its clarity. Today, the solvent used by almost 90% of all dry cleaners is perchloroethylene, commonly known as "perc." It is completely non-flammable and non-combustible, of relatively low toxicity, and can be efficiently reused and recycled.

Dry cleaning is any cleaning process for clothing and textiles using an organic solvent other than water. Dry cleaning is necessary for cleaning items which would otherwise be damaged by water and detergent. Dry cleaning uses non-water-based solvents to remove dirt and stains from clothes. The potential for using petroleum based solvents in this manner was first discovered in the mid-19th century by French dye-works owner Jean Baptiste Jolly, who noticed that his tablecloth became cleaner after his maid spilled kerosene on it, and from this observation developed a service to clean other people's clothes in this manner, which he termed "nettoyage a sec," or "dry cleaning" in English.

Early dry cleaners used petroleum-based solvents, such as gasoline and kerosene. Concerns over flammability led William Joseph Stoddard, a dry cleaner from Atlanta, to develop Stoddard solvent as a slightly less flammable alternative to gasoline-based solvents. The use of highly flammable petroleum solvents led to many fires and explosions, which resulted in heavy regulation of dry cleaners. After World War I, dry cleaners began using various chlorinated solvents. These solvents were much less flammable than petroleum solvents and had much greater cleaning power. By the mid-1930s the dry cleaning industry had adopted tetrachloroethylene (perchloroethylene) as a standard, colloquially called "perc," as the ideal solvent. It is stable, nonflammable, and has excellent cleaning power and is gentle to most garments.

A dry cleaning machine is somewhat similar to combination of a domestic washing machine, and clothes dryer. Garments are placed into a washing/ extraction chamber (referred to as the basket, or drum). This is the core of the dry cleaning machine. The washing chamber contains a horizontal, perforated drum that rotates within an outer shell. The shell holds the solvent while the rotating drum holds the garment load. Depending on the size of the machine the basket capacity will be between 20 and 80 lb (9-36 kg) of garments.

During the wash cycle the chamber is filled approximately 1/3 full of solvent and begins to rotate to agitate the clothing. The solvent temperature is controlled at 85°F (29.4°C), as a higher temperature may extract dye from the garments, causing color loss. During the wash cycle, the chamber is constantly fed a supply of fresh solvent from the working solvent tank while spent solvent is removed and sent to a filter unit comprising a distillation boiler and condenser. The ideal flow rate is one gallon of solvent per pound of garments per minute, depending on the size of the machine. Before being placed in the machine, garments are inspected for stains and soils by the operator.

Depending on the nature of the soil, a catalyst may be applied to the soil; this depends on the operator's judgement of the makeup of the textile and the soil itself. Oil-based soils such as grease, oil or lipstick typically are removed very well by perchloroethylene, whereas water-based soils such as coffee, wine and perspiration will need a catalyst to allow the dry cleaning solvent to emulsify and lift them. Food based grease soils fall in between the two, and a milder catalyst may be applied.

Garments are also checked carefully for foreign objects; items such as plastic pens will melt in the solvent bath and may damage textiles beyond recovery. Some textile dyes are "loose" (red being the main culprit), and will shed dye during the solvent immersion; these will not be included in a load along with lighter-based coloured textiles to avoid colour transfer. In addition the solvent used has been distilled to remove any impurities which may transfer to the clothes.

Garments are checked carefully for dry cleaning textile compatibility, including the fasteners; many decorative fasteners are either not dry cleaning solvent proof, or will not withstand the mechanical action of the cleaning cycle. These will be removed and restiched after the cleaning, or protected with a small padded protector as appropriate. Finally, fragile items such as feather bedspreads or tasselled rugs or hangings may be enclosed into a loose mesh bag. The density of perchloroethylene is around 1.7 kg/L at room temperature (70% heavier than water), and the sheer weight of absorbed solvent may cause the textile to fail under centrifugal force during the extraction cycle unless the mesh bag provides mechanical support. A typical wash cycle lasts for 8-15 minutes depending on the type of garments and amount of soiling.

During the first three minutes, solvent-soluble soils dissolve into the perchloroethylene and loose insoluble soil from fabrics comes off. It takes approximately ten to twelve minutes after the loose soil has come off to remove all of the ground-in insoluble soil from the garments. Machines using hydrocarbon solvents require a much longer wash cycle of at least 25 minutes because of the much slower rate of solvation of solvent soluble soils (e.g oily stains). A drycleaning surfactant "soap" may also be added.

At the end of wash cycle, the machine starts a rinse cycle and the garment load is rinsed with fresh distilled solvent from the pure solvent tank. This pure solvent rinse prevents discoloration of garments caused by soil particles being absorbed back onto the garment surface from the "dirty" working solvent.

After the rinse cycle the machine begins the extraction process. This process recovers dry cleaning solvent for reuse. Modern dry cleaning machines can recover approximately 99.99% of the solvent used in the cleaning process.

The extraction begins by draining the solvent out of the washing chamber cycle and accelerating the basket to speeds of 350 to 450 rpm, causing much of the solvent to spin free of the fabric. When no more solvent can be spun out, the machine starts its drying cycle.

During the drying cycle the garments are tumbled in continuous stream of warm air (145°F/63°C) that circulates through the basket evaporating any traces of solvent left behind after the spin cycle. The temperature of the air is carefully controlled to prevent over drying and heat damage to the garments. The warm air then passes through a chiller unit where the solvent vapors are condensed, and returned to the distilled solvent tank. Modern dry cleaning machines use a closed loop system where the chilled air is then reheated and recirculated. This results in very high solvent recovery rates.

After the drying cycle is completed, a deodorizing (aeration) cycle starts to cool the garments and remove the last traces of dry cleaning solvent, by circulating cool outside air over the garments and then through a vapor recovery filter made from activated carbon and polymer resins. At the end of the aeration cycle, the dry cleaned garments are clean, odorless and ready for pressing/ finishing.

Solvent Processing

Working solvent from the washing chamber passes through several filtration steps before it is returned to the washing chamber. The first step is a button trap which prevents small objects (lint, fasteners, buttons, coins etc) from entering the solvent pump. Next the solvent passes through a filter unit which removes lint and insoluble suspended soils from the solvent. Several different types are used, most filters use an ultra fine mesh to support a thin layer of filter powder (made from diatomaceous earth and activated clays). Some machines use powderless filters which are capable of removing soil particles greater than 30 micrometers from the solvent.

As the machine is used, a thin layer of filter cake (called muck) accumulates on the surface of the lint filter. The muck is removed regularly (once per day) and then further processed to recover any solvent trapped in the muck. Many machines use "spin disc filters" in which the muck is removed from the filter surface by centrifugal action while the filter is back-washed with solvent.

After passing through the lint filter, the solvent passes through an absorptive cartridge filter, this filter is made from activated clays and charcoal and removes fine insoluble soil and non-volatile residues along with dyes from solvent. Finally the solvent passes through a polishing filter which removes any traces of soil not removed by the previous filters. The clean working solvent is then returned to the working solvent tank.

To enhance cleaning power, small amounts of detergent (0.5%-1.5%) are added to the working solvent and are essential to its functionality. These detergents help dissolve hydrophilic soils and keep soil from redepositing on garments. Depending on the machine's design, either an anionic or cationic detergent is used.

Dry Cleaning Wastes

Cooked Muck

Cooked Powder Residue — the waste material generated by cooking down or distilling muck. Cooked powder residue is a hazardous waste and will contain solvent, powdered filter material (diatomite), carbon, non-volatile residues, lint, dyes, grease, soils and water. This material should then be disposed of in accordance with local laws.

Sludge

The waste sludge or solid residue from the still. Still bottoms contain solvent, water, soils, carbon and other non-volatile residues. Still bottoms from chlorinated solvent dry cleaning operations are hazardous wastes.

Environment

Perchloroethylene is toxic, and some believe that long-term exposure can cause liver and kidney damage, though no study has conclusively proven that. There are other solvents including:

- Wet Cleaning — This is a system that uses water and biodegradable soap. Computer-controlled dryers and stretching machines ensure that the fabric retains its natural size and shape. Wet cleaning is claimed to clean a majority of "dry clean only" garments safely, including leather; suede; most tailored woolens, silks and rayons. (Neckties seem to be the one exception.)

- Silicone and Liquid CO2 Solvents — Relatively new approaches to dry cleaning have been developed based on both liquid carbon dioxide and silicone. Dry cleaners using these solvents are currently few in number.
- Alternative Petroleum Solvents — This is more like standard dry cleaning, but the processes use alternative hydrocarbon solvents such as Exxon DF-2000 or Chevron-Phillips' EcoSolv. Alternative solvents tend to be much less effective than perc and very flammable, creating a substantial explosion and fire hazard.

Solvents Used

Modern

- Perchloroethylene — Most common solvent, known more commonly as "Perc", the "standard" for cleaning performance, and most aggressive cleaner (use caution, may cause color bleeding/loss, especially at higher temperatures).
- High flash point hydrocarbons DF-2000 (140°F/60°C flash point) — Slightly less flammable and explosive than Stoddard solvent, not as effective as perchloroethylene.
- Modified hydrocarbons blends (Pure Dry)
- Glycol ethers (dipropylene glycol tertiary-butyl ether) (Rynex) — not as effective as perchloroethylene.
- Cyclic Silicone decamethylcyclopentasiloxane, GreenEarth — not as aggressive as perchloroethylene. Gentler cleaning through lowest surface tension of current solvents. Cost per gallon is similar to perc in most states with environmental surcharges. Solvent mileage higher than perc and hydrocarbon products. Requires a license be obtained to utilize the intellectual property of GreenEarth Cleaning. Degrades within days in the environment to SiO_2 (sand), CO_2 (carbon dioxide), and H_2O (water).
- Liquid CO_2 — Consumer Reports rated this method superior to even conventional methods; machinery is very expensive. Most cleaners with these machines keep traditional machines on-site for the heavier soiled textiles. Not good for use on smoke or water damaged textiles.

Historical

- Stoddard solvent — Very flammable and explosive, 100°F/38°C flash point.

- Carbon tetrachloride — Toxic and corrosive
- Trichloroethane — Overly aggressive and harsh
- Valclene 113 Freon-113 — Ozone destroying CFC

Types

There are two types of garment cleaning: wet cleaning and dry cleaning. Dry cleaning is suitable for delicate fabrics and other clothes which cannot be washed by water. Dry cleaning is a method to wash dirt and stains by using a solvent instead of water. This is a very effective way to remove fatty stains, without changing the garment shape and altering the texture and the color of the fabric materials. Dry cleaning is less effective for washing dirts that are easily soluble in water.

Wet Cleaning

This is a system that uses biodegradable soap and water. Computer-controlled dryers and stretching machines ensure that the fabric retains its natural size and shape. Wet cleaning can purportedly clean 99.9% of "dry clean only" garments safely, including leather; suede; most tailored woolens, silks and rayons. (Neckties seem to be the one exception.)

Wet cleaning starts with the pretreatment of spots and stains using special cleaning agents. Wet cleaning is the process of removing soils from garments and other textile items through the use of water and additives (such as detergent) and using precautions to prevent shrinkage, loss of colour, and fabric distortion.

Special detergents, additives, and finishes sets commercial laundering apart from home laundering. This process enables your cleaner to offer consistent quality shirts at reasonable prices.

Silicone and Liquid CO2 Solvents

Relatively new approaches to dry cleaning have been developed based on both liquid carbon dioxide (usually obtained as a recycled byproduct of other industrial processes) and silicone. Dry cleaners using these solvents are currently few in number.

Alternative Petroleum Solvents

This is more like standard dry cleaning, but the processes use alternative hydrocarbon solvents such as Exxon D-2000 or Chevron-Phillips' EcoSolv. Members of the Grinning Planet team have used dry cleaners that employ both solvents and the results were excellent.

Make the Following Checks When Sending Clothes to a Dry Cleaner:

1. Read care instruction labels of clothes. Check garment materials and any handle-with-care signs. If these are made of materials which

cannot be washed in water, they have to be sent to a dry cleaner.

2. If there are any stains, mark them by rough stitches with a white thread. If you know what caused the stains, you should tell the cleaner about them. In the case of two-piece or three-piece suits, they should be sent together to the cleaner for best results.
3. Empty all pockets.
4. If you find an open seam or a button missing, mend them yourself or ask the cleaner to do so. In the case special buttons are used, take these off the clothes before sending them to the cleaner.
5. Keep the receipt from the cleaner.
6. When cleaned clothes are returned, check the finish thoroughly. If you find the cleaning to be unsatisfactory or find any damage, inform the cleaner promptly.
7. Remove the plastic film cover from the clothes, air them in shade for a while before putting them in the closet.

THE MOST COMMON DRY CLEANING TROUBLES

In 1997, the Laundry and Drycleaning Association received 558 complaints from the public regarding damages made to their clothes in cleaning. The breakdown of the damage complaints and the parties who were found responsible are indicative.

Among damages for which the clients were responsible, the most common was the "change of color" amounting to 66 % of the total cases. Cases of stains made by spilled bleach or damages caused by inappropriate storage are on the increase recently at individual homes. It is important that the consumers also have the basic knowledge about how to clean and keep clothes in good shape.

Professional Cleaning Steps

Professional cleaning involves many different operations, all performed by skilled people and designed to give your garments a fresh and clean appearance. Procedures include:

- Checking the care label for instructions and fiber content
- Classifying the garment according to fabric type, color and degree of soiling
- Removing spots and stains using special equipment and stain removal techniques
- Dry cleaning, wet cleaning, or laundering, only if so labeled
- Reapplying any sizing, water repellency or other finishes when necessary
- Finishing the garment on professional pressing equipment
- Replacing buttons or performing minor repairs as necessary
- Packaging the garment in protective wrapping

The Importance of Care Labels

The Federal Trade Commission (FTC) requires that manufacturers attach a permanent care label to textile garments to provide directions for their care. Manufacturers must list at least one method of safe care for a garment. The rule covers all textile clothing except footwear, gloves, hats, suede and leather clothing, and household items such as linens. The rule stipulates that the care label is easily found, will not separate from the garment, and will remain legible for the garment's useful life.

The label must warn about any part of the recommended care method that would harm the garment of other garments cleaned with it. It must also warn when there is no method for cleaning a garment without damaging it. Symbols also may appear on a care label to supplement written instructions.

If a label indicates dry cleaning, all components of the garment, including the outer shell, lining, buttons, interfacing, fusing material, and trim, should be colorfast and remain unaltered during cleaning. If any such problem occurs, it is the responsibility of the manufacturer. If you or your cleaner follow the manufacturer's instructions and the garment is damaged, you should return the garment to the store and explain what happened.

If the store will not resolve the problem, write to the manufacturer and send a copy of your complaint letter to the FTC. The information you provide the FTC may reveal a pattern or practice requiring the Commission's attention. If you purchase a garment with no care label, you should contact the FTC, giving the name and address of the store and manufacturer.

Occasionally, damage done in dry cleaning is the responsibility of the dry cleaner and not the result of preexisting conditions or defects. In such cases, the cleaner will usually settle the claim promptly and fairly, often using IFI's Fair Claims Guide. If there is some doubt about responsibility, the member cleaner can send the garment to IFI's Garment Analysis Laboratory to determine the cause of the problem.

Help Your Clothes and Your Cleaner

- Bring your garments in for professional cleaning as soon as possible after staining occurs. Stains or soils left too long may become permanent.
- Discuss any stains with your cleaner.
- Keep perfumes, lotions, deodorants, antiperspirants, and other toiletries away from your clothes. These products likely contain alcohol which will damage some dyes.
- Protect garment, especially those made of silk, from excessive perspiration, as this can cause dyes to discolour.
- Have matching pieces of an ensemble cleaned together so any color fading will be uniform.

- Protect your garments from prolonged exposure to direct light.
- Don't press stained or soiled clothing, as the heat may set some stains.

IRONING PROCESS

Ironing or smoothing is the work of using a heated tool to remove wrinkles from washed clothes. The common tools for this purpose are called "irons", though modern designs are no longer made of iron.

Ironing works by loosening the bonds between the long-chain polymer molecules in the fibres of the material. While the molecules are hot, the fibres are straightened by the weight of the iron, and they hold their new shape as they cool. Some fabrics, such as cotton, require the addition of water to loosen the intermolecular bonds. Many modern fabrics (developed in or after the mid-twentieth century) are advertised as needing little or no ironing.

Ironing may also be used as a germ/parasite killing hygienic operation. Metal pans filled with charcoal were used for smoothing fabrics in China in the 1st century BC. From the 17th century, sadirons or sad irons (from an old word meaning solid) began to be used. They were thick slabs of cast iron, delta-shaped and with a handle, heated in a fire. A later design consisted of an iron box which could be filled with hot coals, which had to be periodically aerated by attaching a bellows. This type was on sale in the USA until at least 1902.

In the late nineteenth and early twentieth centuries, there were many irons in use which were heated by a fuel such as kerosene, alcohol, whale oil, natural gas, carbide gas (acetylene) as with carbide lamps, or even gasoline. Some houses were equipped with a system of pipes for distributing natural gas or carbide gas to different rooms in order to operate appliances such as irons, in addition to lights. Despite the risk of fire, liquid-fuel irons were sold in U.S. rural areas up through World War II. On 16 February 1858 W. Vandenburg, and J. Harvey, patented an ironing board that made pressing sleeves and pant legs easier (U.S. Patent 19,390).

In the industrialized world, these designs have been superseded by the electric iron, which uses resistive heating from an electric current. The hot plate, called the sole plate, is no longer made of iron but of aluminium or stainless steel. The heating element is controlled by a thermostat which switches the current on and off to maintain the selected temperature.

The invention of the resistively heated electric iron is credited to Henry W. Seeley of New York in 1882. In the same year an iron heated by a carbon arc was introduced in France, but was too dangerous to be successful. The early electric irons had no easy way to control their temperature, and the first thermostatically controlled electric iron appeared in the 1920s. Later, steam was used to iron clothing. Credit to the invention of the steam iron goes to Thomas Sears.

Modern Irons

Irons are normally classified into automatic and non automatic irons. Modern irons available for sale to consumers have some or all of the following features (more expensive models have more features, as one would expect):

- Ejection of steam through the clothing during the ironing process;
- A water reservoir inside the iron (used for the steam generation);
- An indicator showing the amount of water left in the reservoir;
- A thermostat ensuring maintenance of a constant temperature;
- Convenient method for setting the iron down, usually on its end/ vertically, so as to not allow contact between the hot part and either table or clothes;
- A temperature control dial showing a range of possible temperatures (typically marked with types of cloth instead of absolute degree measurements, for example, "wool", "cotton", "linen", etc.);
- Constant steam feature - constantly sends steam through the hot part of the iron into the clothes;
- Cord control - the point at which the cord attaches to the iron has a spring to hold the cord out of the way while ironing and likewise when setting down the iron (prevents fires, is more convenient, etc.).
- Steam burst feature - sends a burst of steam through the clothes when the user presses a button;
- (advanced feature) Dial controlling the amount of steam to emit as a constant stream;
- (advanced feature) Anti-burn control - if the iron is left flat (hot part touching clothes) for too long, the iron shuts off to prevent fires;
- (advanced feature) Energy saving control - if the iron is left undisturbed for several (10 or 15) minutes, the iron shuts off to save energy and prevent fires.
- Cordless irons - the iron is placed on a stand for a short period to warm up, using thermal mass to stay hot for a short period. These are useful for light loads only. (Unlike other cordless appliances, cordless irons are not powered from rechargeable batteries, because it would be difficult for batteries to provide enough power to heat the iron.)
- (advanced feature) 3 way auto shut off
- (advanced feature) self-cleaning
- (advanced feature) Anti-drip system
  Robots and major home appliances
- Automatic ironing robots, like the Siemens AG Dressman (see link below), have recently appeared on the market.
- Automatic ironing (major) home appliances: Driron.

Usage Notes

- Irons are commonly available as consumer goods.
- Some commercial-grade irons have a boiler unit separate from the handheld iron.
- Most ironing is done on an ironing board, a small, portable, foldable table with a heat resistant top. [African American, Sarah Boone, patented an improvement to the ironing board on April 26, 1892. Sarah Boone's ironing board was designed to be effective in ironing the sleeves and bodies of ladies' garments. Sarah Boone's board was very narrow and curved, the size and fit of a sleeve, and it was reversible, making it easy to iron both sides of a sleeve.]
- Some commercial-grade ironing boards incorporate a heating element and a pedal-operated vacuum to pull air through the board and dry the garment.
- Permanent press clothing was developed to reduce the ironing necessary by combining wrinkle-resistant polyester with cotton.
- Commercial laundries typically use steam presses to iron clothes instead of irons;
- Clothes such as shirts, trousers, and skirts are typically ironed, while underwear, socks, sheets, sweaters, and materials where wrinkling is not a factor are not, although this depends on the culture and circumstances.
- Irons cause many fires and injuries each year due to their being very heavy, very hot, and often used on ironing boards that can fall over onto small children.
- Irons used to be much more dangerous due to the electrical cords fraying; modern insulation is much safer and better able to cope with being heated.

Safe Ironing Temperatures

- Acetate: 143° C (290° F)
- Acrylic: 135° C (275° F)
- Triacetate: 200° C (390° F)
- Cotton: 204° C (400° F)
- Linen: 230° C (445° F)
- Lycra/Spandex: 135° C (275° F)
- Nylon 6.6: 135° C (275° F)
- Polyester: 148° C (300° F)
- Silk: 148 °C (300° F)
- Viscose: 190° C (375° F)
- Wool: 148° C (300° F)

Continuous manual ironing can be a cause of repetitive strain injury to the user’s wrist. There are still a few of us who like to iron and feel our clothes look their best when time and care are taken to iron. Sure there are a lot of wash and wear fabrics, permanent press, new finishs and most are not ironing, but even if they say wash and wear the quality is not there unless they are ironed, they just look better. When you buy an iron or replace the one you have try to buy the best you can as it will be worth the extra expense to obtain.

Today's dry, steam and spray-stream irons are designed to control the correct heat for the many different types of fabrics in your home. If you will look on the back of the labels on your clothing you'll see ironing instructions or let's hope you will. Check the directions for your individual iron to know what setting to use for the particular material you plan to iron.

And remember you don't need to apply pressure just glide the iron along the material. You will also need to purchase a good ironing board, one that will not tip over or collapse as you iron. Look for one with a sturdy metal top with adequate ventilation for the steam to penetrate, and you'll want the board to be lightweight also and adjustable as some of us are tall and others are shorter. Look around and check to see if you can purchase a sleeve to attach to the ironing board.

Also some can be lowered enough that you could sit in a chair and iron. You'll also need to have the ironing board near a good light, such as you would use for reading as you'll need to be able to see what you are ironing to get the best possible ironing done.

First of all, sort everything you plan to iron in one day. If you will start ironing the articles requiring the lowest steam heat setting, you won't have to wait for the iron to cool before ironing other articles. Don't forget when you finish the steam ironing to empty your iron and switch to the dry settings and higher temperatures.

Begin by trying the setting on the article as you can always turn the high higher, gradually increase the heat as needed. Straighten your material on the ironing board, use the palms of your hands and smooth from the center outward, try to iron with straight strokes.

Iron each section completely before you start on the next. Iron the collars, cuffs, sleeves, belts, and trimmings first then the flat sections of the garments. Also remember that white and light colored cottons and linens are ironed on the right side, dark cottons and linens on the wrong side; silks and rayons on the wrong side, damask first on the right side, then ironed on the wrong side.

Make sure the iron is dry before putting it away also. Here are a few ironing tips that I might add to help. Before ironing handkerchiefs, fold them in half and pull tight. Lay flat on the ironing board and iron up from the fold. This will prevent uneven edges. Continue ironing and fold as desired. If spray starch sticks to the soleplate or bottom of the iron, let the iron cool and apply paste silver polish, then wipe with a damp cloth and dry. Or you can remove rough and sticky spots from the iron by sprinkling a little salt on a piece of paper or cloth and running the hot iron over it, but don't do this on irons with teflon soleplates.

Using Starch

When you iron with starch, make sure you iron quickly. Otherwise, the starch may stick to the sole plate of the iron and burn. When using spray starch, spray the entire garment, then roll it up and allow it to sit for one minute to let the starch penetrate the fabric. This will prevent starch from building on the sole plate of your iron.

Go with the grain

Always press with the grain of the fabric to prevent it from stretching.

Avoid shine marks.

By ironing clothes on the wrong side, or use a pressing cloth on the right side to avoid shine marks.

Comfort

The ironing board should be at hip level to prevent back strain when ironing laundry. Choose an ironing board with adjustable height options so you won't have to bend over. Keep the iron moving to prevent scorching or overheating. Pressing too hard can damage a fabric's finish. However, pressing the iron's tip into the middle of a seam helps to open and flatten it out.

Testing Fabrics First

The surface of many dark-colored garments can take on a shine after being ironed, particularly on seams or hems. To check how the garment will react, turn it inside out and press one of these thicker areas where it's not likely to be seen. If the iron feels like it is sticking to the fabric, or a shine is apparent, be careful in proceeding. You may have to use a very low heat setting and a very light touch to remove wrinkles. Also try pressing the garment from the inside to avoid leaving marks on the fabric. Wools, tweeds, and corduroys, on the other hand, take nicely to a dampened pressing cloth, which adds moisture and protects the fabric finish

Wrinkling begins in the washer and dryer

Wash bright colors and lightly soiled fabrics in cold water to minimize washer wrinkling. Don't overload the dryer, as it prevents clothes from tumbling properly and causes more wrinkles. Don't overdry clothes, as excessive heat can set wrinkles.

After Ironing

After ironing clothes, let garments hang until they are cool and dry before wearing. This sets the press, keeping clothing smooth and crisp much longer.

Ironing Silk

Iron silk inside-out on a low heat setting.
Ironing wool, mohair, cashmere, camel, alpaca.
Use steam and a medium heat setting.

Rayon

Iron inside-out on low heat.

Polyester

Iron on a low or medium heat setting.

Nylon

Iron on a low heat setting.

Choosing an Iron

This section of our ironing help articles deals with selecting an iron. When looking for the best clothes iron, the most important thing is to consider your ironing needs. This page contains a great deal of information on clothes iron features, which should help you determine what is available and what you need. Most clothes iron manufacturers use different terminology for the same features, so this article should also help you better understand what their names mean.

- Automatic safety shut off feature. Most clothes irons have this feature, but not all do. There are also variances in how long the irons shut off when not in use. Most people will want this feature, but if you are a crafter or do a lot of sewing enthusiast, nothing is more annoying than having your iron shut off every 10 minutes.
- How long does the iron take to warm up? How long does it take to re-heat?
- Does it have adjustable heat settings, from delicate fabrics to cotton/ wool?
- Is the iron cordless or not? If it has a cord, how long is it? Reversible or pivoting cords allow right and left handed ironers to work comfortably.
- Does the iron offer a steam setting, or not? Does it offer a burst of steam? A burst of steam is particularly helpful for eliminating tough wrinkles.
- Does it have a non-stick sole plate? Non-stick soleplates are usually

the same type of coating material used on cookware. A soleplate that cleans easily is a plus.

- Does it have a self-clean feature? Some types of clothes iron use a burst of steam to clear the holes in the sole plate.
- What is the size of the water tank? The bigger the water tank, the less frequently you will need to refill it. Also, some irons have removable water tanks, so you don't have to unplug the iron when refilling. A clear gauge on the water tank allows you to visually monitor the water level.
- Vertical steam allows you to use the iron like a steamer, to steam out wrinkles on things like curtains and hanging garments.
- An anti-calcium filter helps prevent mineral buildup, which can clog steam vents. Most higher end models have this features, but other types of clothes iron may too.
- Most irons these days do not require you to use distilled water. In fact, distilled water may even harm your clothes iron, as may softened water. As a result, you will definitely want to check the instructions before putting any water into the clothes iron.
- Feel. The best clothes iron for you will have a contoured handle and built-in thumb rest that feels comfortable in your hand when you iron laundry, and isn't too heavy to lift or move.
- Variable steam. Variable steam automatically adjusts the amount of steam for the heat setting, ensuring you don't get high steam on more delicate fabrics.
- Weight. A full size iron weighs about 3 lbs, whereas a lightweight one weighs about 1.5 lbs.

PROCESS OF BLEACHING

To bleach something is to remove or lighten its color; a "bleach" is a chemical that can produce these effects, often via oxidation. Common chemical bleaches include a solution of sodium hypochlorite (NaClO), or "chlorine bleach," and "oxygen bleach," which contains hydrogen peroxide or a peroxide-releasing compound such as sodium perborate or sodium percarbonate. "Bleaching powder" is calcium hypochlorite. Bleaching may be a preliminary step in the process of dyeing.

Types of Bleach

Household bleach, also known as chlorine bleach, sodium hypochlorite (NaClO), has a pH level of 11 and is used in the home for whitening clothes, removing stains, and disinfecting. This is because sodium hypochlorite yields chlorine radicals — oxidizing agents readily reacting with many substances. Using chlorine bleach on garments made of wool, nylon, silk, leather or any amount of spandex will stain the garment yellow which is permanent or very difficult to remove.

Chlorine Bleach

Chlorine bleach is often used with laundry detergents and is also commonly used as a disinfectant. Mixing bleach and cleaners containing ammonia, or using bleach to clean up urine can create toxic chloramine gases and an explosive called nitrogen trichloride.

Chlorine bleach comes in liquid and powdered forms. A variety of house and name brands are available. All liquid bleaches contain approximately 5% active ingredient. Select the product that provides the best results.

Liquid chlorine bleaches are usually less expensive than powdered forms. They kill germs, too. Some fabrics and finishes are damaged or discolored by chlorine bleaches, so read the labels on clothes and bleach containers before using chlorine bleach. Iron deposits in water will interact with chlorine bleach and make rust stains on clothes. Chlorine bleach also makes some yellow stains worse and may make holes in fabric.

Chlorine bleaches may be used on:

Dish towels Bath towels

Sheets Color-fast prints

Pillowcases Tablecloths

White socks T-shirts and undershirts

Other chlorine bleach-safe items

Do not use chlorine bleach on silk, wool, spandex stretch fabrics, blends of these fibers, or fabrics with certain dyes and finishes. Check clothes labels and hang-tags to determine if they can be safely bleached. For most top-loading washing machines, do not use more than 1 cup of liquid chlorine bleach. If the machine is extra large, use 1-1/4 cup of bleach. For front-loading machines, do not use more than 1/2 cup liquid bleach. For hand washing, do not use more than 1 tablespoon of bleach per gallon of water.

Mix the bleach with lots of water before mixing with soapy wash water. Otherwise the bleach may make holes in the fabrics. Rinse clothes well to remove the bleach, as it may weaken the clothes if it stays in them.

Oxygen Bleach

Most oxygen bleaches are powders. They can be safely used on any clothes that need bleaching and are safe for all fabrics, fabric finishes, and colored fabrics. Look for the words, "Safe for all fabrics" on the package. A variety of oxygen bleach products are available.

- Add oxygen bleach to the wash water.
- Most oxygen bleaches work best in hot water.
- Follow directions on the package.

- Rinse the clothes well.
- For best results, use oxygen bleach each time the items are washed.

FABRIC SOFTENERS

Fabric softeners and water softeners are not the same. Fabric softeners make clothes soft and fluffy and reduce static electricity that makes clothes cling. Water softeners, however, remove mineral deposits from hard water.

Fabric softeners come as liquids, dryer sheets, and in dryer dispensers. Liquid fabric softeners are added to the final rinse water of the washing cycle. Fabric softener sheets are added to each drying cycle. Fabric softener dispensers are fixed to the side of the dryer and soften laundry for several drying cycles before they must be replaced.

- Follow the package instructions carefully for best results.
- To prevent spotting, be sure to dilute liquid softener in water before adding it to the laundry.
- Dryer sheets may cause oily spots on some fabrics.
- Check garments carefully as they are removed from the dryer. If spots occur, rub them with a detergent paste and rewash.

SORTING AND INSPECTING LAUNDRY

Sort and check all clothes before putting them into the washer. This will help keep clothes looking nice longer. Choosing the right water temperature for different fabrics and treating stains before laundering also keep clothes looking nice.

- Test colored clothes to see if the colors run. If they do, wash them separately or with other clothes of the same color.
- To keep white clothes bright, don't wash them with colored clothes.
- Do not wash lightly soiled clothes with those that are heavily soiled.
- Sort clothes that shed lint from those that don't.
- Separate delicate items from sturdy ones.
- Don't overload the washer. Each load of clothes should have enough room for good washing action.
- Mix small items with large ones for good washing action. For example, put wash cloths and towels in with sheets.

Inspect Every Item

- Empty and brush out pockets and cuffs.
- Mend small rips and tears.
- Remove pins and ornaments.
- Close zippers and hook fasteners.
- Check for stains that may need special attention.

Pretreating and Removing Stains

Before washing, rub heavily soiled areas such as shirt collars or cuffs with a wet bar of soap, liquid detergent, or a paste made with powdered detergent and water. This will loosen heavy soil. On permanent press clothes, pretreat grease spots with soap or detergent and let stand for at least two hours. The sooner a stain is treated, the easier it is to remove. These stains can usually be removed by pretreating with soap or detergent:

Cosmetic Crayon Grease

Ice Cream Oil Grass

Milk Light scorch

If the clothes are still stained after they have been washed and rinsed, try using a bleach. Read the label on the bleach container to be sure it is safe to use on the stained clothes. If fresh blood, egg, catsup, or soft drinks stain washable clothes, soak the stained spot in cool water for half an hour or more. Then rub gently. If the stain doesn't come out, work soap or detergent into the cloth. Then rinse.

Do the same for chocolate, soup, and gravy stains. Let the clothing dry. If a greasy stain is still there, sponge gently with a non-flammable, dry cleaning fluid, which can be purchased at drug or grocery stores. Follow directions on the label carefully. Coffee, tea, and fruit juice stains will usually come out after soaking and rubbing. However, if boiling water won't hurt the cloth, there's a faster and easier way to remove these stains. Hold the cloth over a sink or bucket and pour boiling water through the spot. Pouring the water from about two feet above the spot will force the stain out.

ELIMINATING GERMS FROM CLOTHES

Boiling clothes or hanging them in the sunshine were once common ways to get rid of germs. With modern fabrics, fabric finishes, and washing methods, other ways can be used to disinfect the family laundry. Liquid chlorine bleach that contains 5.25% sodium hypochlorite will destroy most germs. Quaternary, pine oil, or phenolic disinfectants can be safely used on any washable fabrics. Pine oil disinfectants must contain at least 80% pine oil to destroy germs.

It is always a good idea to use a disinfectant in the wash when someone in the family has a bad cold, the flu, or some other infectious illness. Disinfecting the wash will help prevent other family members from getting the same illness.

Use a disinfectant when washing at a laundromat. Illness from another family can be passed on if the washing machine is not disinfected before it is used. Wipe off the surface of the machine with a disinfectant, then add disinfectant to the wash cycle. Follow the directions on the disinfectant's label. Hair bleach contains H_2O_2 (hydrogen peroxide), which gives off oxygen radicals as it decomposes. Oxygen and chlorine radicals both have comparable bleaching effects.

Various other peroxide yielding chemicals are used as bleaching additives. Sodium perborate, sodium percarbonate, sodium persulfate, sodium perphosphate, sodium persilicate, their ammonium, potassium and lithium analogs, calcium peroxide, zinc peroxide, sodium peroxide, carbamide peroxide, and others are commonly used in detergents, toothpastes, and other products.

Chlorine dioxide is used for the bleaching of wood pulp, fats and oils, cellulose, flour, textiles, beeswax, and in a number of other industries. In the food industry, some organic peroxides (benzoyl peroxide, etc.) and other agents (e.g. bromates) are used as flour bleaching and maturing agents. Not all bleaches have to be of oxidizing nature. Sodium dithionite is used as a powerful reducing agent in some bleaching formulas.

How Bleaches Work

Colour in most dyes and pigments are produced by molecules, such as beta carotene, that contain moieties (pieces) known as chromophores. Chemical bleaches work in one of two ways:

- An oxidizing bleach works by breaking the chemical bonds that make up the chromophore. This changes the molecule into a different substance that either does not contain a chromophore, or contains a chromophore that does not absorb visible light.
- A reducing bleach works by converting double bonds in the chromophore into single bonds. This eliminates the ability of the chromophore to absorb visible light.

Sunlight acts as a bleach through a process leading to similar results: high energy photons of light, often in the violet or ultraviolet range, can disrupt the bonds in the chromophore, rendering the resulting substance colorless.

Hazards

A problem with chlorine is that it reacts with organic material to form trihalomethanes like chloroform, which is a well known carcinogen. There is debate over whether any risk from the chloroform in treated drinking water is worth the benefits. However, the use of elemental chlorine in industrial processes such as paper bleaching, with its attendant production of organochlorine-persistent organic pollutants (including dioxins), does not have any benefits. As a consequence over 80 % of the woodpulp is nowadays bleached with chlorine dioxide, reducing the dioxin generation under detectable levels.

Chlorine is a respiratory irritant. It also attacks mucous membranes and burns the skin. As little as 3.5 ppm can be detected as an odor, and 1000 ppm is likely to be fatal after a few deep breaths. Exposure to chlorine should not exceed 0.5 ppm (8-hour time-weighted average - 40 hour week). Another hazard is the formation of acrid chloramine fumes when hypochlorite bleach comes into contact with ammonia or urine, which, though not nearly as dangerous as chlorine, can cause severe respiratory distress. For these reasons, some consumers prefer the use of natural cleaning products as an alternative to chemical cleaners.

Chlorine was first characterized by the Swedish chemist Carl Wilhelm Scheele in 1774. (As an adherent of the Phlogiston theory, he called it "dephlogisticated marine acid".) French chemist Claude Louis Berthollet, noting the bleaching properties of chlorine, invented hypochlorite bleach in 1789. In French, bleach is known as Eau de Javel, after the village where it was manufactured.

Bleaching Agents

Bleaching agents are compounds which are used to remove color from substances such as textiles. In earlier times textiles were bleached by exposure to the sun and air. Today most commercial bleaches are oxidizing agents, such as sodium hypochlorite (NaOCl) or hydrogen peroxide (H_2O_2) which are quite effective in "decolorizing" substances via oxidation. The action of these bleaches can be illustrated in the following simplified way:

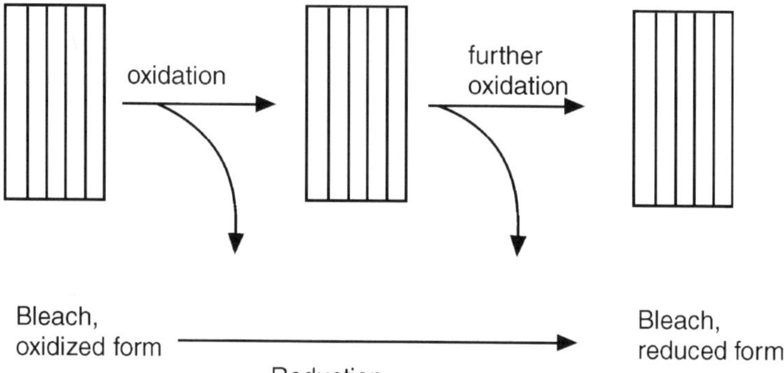

Recall that an oxidizing agent is any substance which causes another substance to lose one or more electrons. The decolorizing action of bleaches is due in part to their ability to remove these electrons which are activated by visible light to produce the various colors. The hypochlorite ion (OCl^-), found in many commercial preparations, is reduced to chloride ions and hydroxide ions forming a basic solution as it accepts electrons from the colored material as shown below.

$$OCl^- + 2e^- + HOH \rightarrow Cl^- + 2\ OH^-$$

Bleaches are often combined with "optical brighteners". These compounds are quite different from bleaches. They are capable of absorbing wavelengths of ultraviolet light invisible to the human eye, and converting these wavelengths to blue or blue-green light. The blue or blue-green light is then reflected by the substance making the fabric appear much "whiter and brighter" as more visible light is seen by the eye.

6

Maintenance and Engineering Technology in Hotel Industry

IMPORTANT MANAGEMENT FUNCTIONS OF STAFFING

Staffing, which is one of a hotel's most important management functions, is an ongoing challenge because of the high rate of employee and manager turnover. Full-service hotels can experience annual turnover rates in excess of 100 per cent in certain employee classifications. Some managers consider an annual employee turnover rate of 33 per cent low. At this rate, the entire hotel must be completely restaffed every three years. The higher the turnover rate, the larger the number of employees who must be replaced.

For example, if a hotel with 450 employees has a 75 per cent annual turnover rate, it will be completely restaffed every 16 months. Staffing is the responsibility of the human resources department. In an attempt to reduce employee turnover, hotel and lodging businesses are giving increasing attention to job design, seeking to enhance those job characteristics that give the employee the greatest satisfaction and motivation. Good job design must take into account the needs of employees as well as the demands of the job. Well-thought-out job design begins when management conducts a job analysis—that is, a thorough evaluation of the specific tasks performed for a particular job and the time required to perform them. Job analysis is an ongoing process, as many jobs change with improvements in technology and pressure to improve product quality. The job analysis is the basis for the job description and job specification.

A job description includes the job title, pay, a brief statement of duties and procedures, working conditions, and hours. The job specification is an outline of the qualifications necessary for a particular job. In response to the limits of specialization, organizations can redesign jobs to improve coordination, productivity, and product quality while responding to an employee's needs for learning, challenge, variety, increased responsibility, and achievement.

Such job redesign often involves job rotation, the systematic movement of employees from one job to another; job enlargement, an increase in the number of tasks an employee will do in the job; job enrichment, the attempt to give the employee more control over job-related activities; and flextime, a flexible work schedule that permits employee input in establishing work schedules. In team-driven job redesign, a concept similar to job rotation, employees can transfer back and forth among teams that provide different services or products.

Hotels recruit employees from a variety of sources. Newspapers and employee referrals are used to recruit nonskilled hourly employees. Supervisory and management employees generally are recruited through colleges and universities, promotions from within, professional associations, and management recruiters. Hotels that take more time in making their selections are more successful in retaining employees. Discussions of employee training and development often concentrate on training techniques without giving a full explanation of what a hotel is trying to accomplish. As training and development impart job skills and educate employees, supervisors, and managers, they also improve current and future employee performance, which affects the bottom line. Effective training includes problem solving, problem analysis, quality measurement and feedback, and team building.

Performance evaluation, also called performance appraisal, is the systematic review of the strengths and weaknesses of an employee's performance. The major difficulty in a performance appraisal is quantifying those strengths and weaknesses. The performance of some jobs is easy to quantify, while for others it is more difficult. An important part of the appraisal process is a well-established job description, so that the employee and the supervisor have similar expectations.

Compensation includes the monetary and nonmonetary rewards that managers, supervisors, and employees receive for performing their jobs. In order to set compensation levels, the human resources department must periodically conduct job evaluations, which determine the value of the job to the hotel. Knowledge of the value of the job to the organization and of wage rates for each job classification allows the hotel to establish a fair compensation policy.

DUTIES AND RESPOSIBILITIES OF HOUSEKEEPING STAFF

Duties and Responsibilities:

- All Ledgers are to be put up for Joint Directors signature through housekeeping Manager on a quarterly basis.
- Daily use of cleaning agents are to be recorded and the records to be put up for housekeeping managers signature monthly.
- Ensure proper discipline among Housekeepers, Sweepers, Dhobi, and Tailor working in linen department. Make sure that they wear name badges.

- Floor scrubbing Machine is to be taken by other housekeeper only with the permission of the linen housekeeper.
- Floor scrubbing machines are to be chain locked when not in use.
- Any other job given by the housekeeping manager or the senior management staff from time to time.
- Linen and curtains are to be checked for wear and tear quarterly.
- Maintain a sepearate Indent for the cleaning agents used in Laundry and the Indent should be signed by the housekeeping Manager.
- Periodical Maintenance of washing Machine, Ironing Machine, Iron Box, Sewing Machine, and floor scrubbing Machine to be carried out and the record to be maintained.
- Planning and scheduling the standard procedure for cleaning all types of linen to be displayed in laundry and to be followed strictly.
- Responsible for the cleanliness of sweepers uniform.
- Sell all waste materials and deposit the sale proceeds to secretary.
- Supervision and checking the cleanliness of the surrounding area of main hospital.
- Tailoring rates for various items of stitched materials are to be maintained. For any revision in charges prior approval to be taken from secretary.
- Time schedule to be prepared for collection of clean/soiled linen and to be followed strictly.
- Torn or faded linen is to be segregated and to be shown to joint director quarterly. If any item is found unserviceable the same should be written off the ledger getting after the approval of Jt.Director.
- Torn or faded linen is to be used for cleaning purposes.
- Train and Guide the trainees housekeeper posted in linen department.
- Training classes are to be taken for junior housekeepers when required.
- Washing charges for various linen are to be fixed and approval. For any change in the charges, prior approval to be taken from secretary.

Designation – Floor Executive: Housekeeping:

- Control over all keys of all departments in OPDS.
- Disposal of refuse is to be monitored.
- Ensure proper discipline of Housekeepers and sweepers working in out patient department and make sure that they wear name badge.
- Experimenting new techniques and equipments for cleaning.
- Any other job given by the housekeeping manager or the senior management staff from time to time.

- Indenting cleaning agents and equipments and ensure their proper use and keeping records.
- Maintain the Leave list of all sweepers working in main hospital, free hospital, Laico and other supporting areas and day off Roster of sweepers working in OPDs.
- Maintaining the checklist of furniture, electrical plumbing, carpentry and mason work in the OPDs.
- Maintenance details of continuous duties of housekeepers.
- Maintenance of House Keeping department room and attending to phone calls promptly.
- Maintenance of sweepers attendance Register, wage sheets, deployment, control and details for continuous duties.
- Maintenance responsible for the sweepers uniform collected after working hours from the sweepers.
- Make sure the tyre cart is secured and locked after working hours.
- Planning and scheduling the standard procedures in writing for the proper care of all types of floors, walls, windows, furniture, bathrooms, Lavatories and other equipments to be prepared and follows trictly.
- Preparing inventories for furniture and keeping records for their maintenance.
- Reporting and checking of all maintenance work.
- Responsible for the cleaning schedule of freshwater tank, filters, coolers and septic Tanks and the cleaning and due dates are to be displayed in the boards as well as in the Register. The register shall be put up for Housekeeping managers signature every month.
- Supervision and checking the cleanliness of out patient departments and Basement area.
- Supervision of quality maintenance and infection control.
- To develop proper use of cleaning procedures.
- Training and taking classes for juniors

Designation – Ward Executive: Housekeeping:

- Attending to patient's complaints.
- Control of all keys of 3rd and 4th floor.
- Development and standardization of newer and improved method of cleaning for better results at lower costs.
- Ensure proper discipline of Housekeepers and sweepers in 3rd and 4th floor and make sure they wear name badges.
- Ensure upkeep and maintenance of hospital property
- Any other job that may be assigned by the housekeeping manager or the senior management staff from time to time.
- Indenting cleaning agents and equipments and ensure their proper use and keeping records.

- Maintain day off roster for sweepers in 3rd and 4th floor.
- Maintaining the checklist of furniture, Electrical, plumbing, Carpenting and Mason works in 3rd and 4th floor.
- Planning and scheduling the standard procedures in writing for the proper care of all types of floors walls, windows, furniture, bathrooms lavatories and other equipments to be placed and to be followed strictly.
- Preparing inventories for 3rd and 4th floor furniture and keeping records for their maintenance.
- Responsible for the sweepers uniform collected after working hours from the sweepers of 3rd and 4th floor.
- Supervision and checking the cleanliness of 3rd and 4th floor.
- Supervision for quality maintenance and infection control.
- Take adequate precautions in controlling cross infections.
- To assist in quality health care and patient care programme
- Training and taking classes for juniors.

Designation: - Ward Executive: Housekeeping:

- Attending to patients complaints
- Control of all keys of 1st and 2nd Floor
- Development and standardization of newer and improved method of cleaning for better results at lower costs.
- Ensure proper discipline of housekeepers and sweepers in 1st and 2nd floor and make sure the wear name badges.
- Ensure up keep and maintenance of hospital property.
- Any other job that may be assigned by the housekeeping Manager or the senior management staff from time to time.
- Indenting cleaning agents and equipments and ensure their proper use and keeping necessary records.
- Maintain day off roaster of sweepers in 1st and 2nd floor.
- Planning and scheduling the standard procedures in writing for the proper care of all types of floors, walls, windows, furniture, bathrooms, lavatories and other equipments to be placed and to be followed strictly.
- Preparing inventories for 1st and 2nd floor furniture and keeping records for their maintenance.
- Responsible for the sweepers uniform collected after working hours from the sweepers of 1st and 2nd floor.
- Supervision and checking the cleanliness of 1st and 2nd floor.
- Supervision for quality maintenance and infection control.
- Take adequate precautions for controlling cross infections.
- To assist in quality health care and patient care programme.
- Training and taking classes for juniors

Designation – Senior Trainees: Housekeeping:

- Attend all theory and practical classes as per the programme.

- Indenting cleaning agents, equipments and ensure the proper use and maintain their records.
- Planning and scheduling the standard procedures for cleaning in writing and follow strictly.
- Preparing inventories for furniture and keeping records for their maintenance.
- Work under the supervision of executive Linen housekeeper.

CLEANING AND CLEANING MATERIALS : PRIMARY ASPECT OF HOUSEKEEPING

We know that any establishment has to be clean, well maintained and presentable at any given moment of time. But how to ensure well maintained premises? Cleaning is the most important and primary aspect of housekeeping. It is a process of removing dirt, dust and grime by using methods such as dusting, shaking, sweeping, mopping, washing or polishing. There are certain areas you may clean daily, whereas you may clean other areas occasionally or once/twice in a year. Since there are different types of surfaces like wall, counter tops, marble floors, ceramic tiles, wooden chairs, etc, special cleaning agents are used to clean these specific surfaces. As suggested, discuss these various aspects of cleaning as well as the materials and equipments used for cleaning.

MEANING AND IMPORTANCE OF CLEANING

Cleaning involves sweeping floors, dusting furniture and other surfaces, mopping or washing floors, polishing surfaces, objects and accessories, scrubbing tiles, sinks, toilets, disinfecting drains, rearranging cleaned areas and putting things in their specific place. We can say that cleaning is a process of removing dust, dirt or any other undesirable materials like stains, spots, contents of an ashtray, etc.

What happens if cleaning is not done on a regular basis:

- Yes, your house will become the breeding ground of insects such as cockroaches, spiders, ants, flies and mosquitoes. It will look dirty and will be most uncomfortable.
- Living in such circumstances can also lead to diseases such as asthma, bronchitis, etc. Thus, cleaning is necessary for a general presentable appearance and also to ensure good hygienic conditions. What do you understand by dust and dirt?
- 'Dust' collectively refers to the loose particles, which are very easily moved by air and settle on any surface. It is easily removed with the help of a dry cloth.
- 'Dirt' refers to dust which sticks to any surface with the help of moisture or grease. It is more difficult to remove dirt as compared to dust. Dirt

has to be removed either with a detergent or any other cleaning agent. Let us now read ahead of some general methods of cleaning.

METHODS OF CLEANING

That dust and dirt can be removed by dusting, mopping etc.
Based on these we can describe the cleaning methods as follows:

- *Dusting*: You are already familiar with the term 'dust'. But how do you remove dust? When any surface is wiped with a piece of dry cloth, it carries the loose dust with it and the process is known as dusting. This should be done with a clean soft cloth.
- *Shaking and Beating*: What happens when you shake a cloth full of dust? Yes, the dust falls out. Similarly when you shake or beat any soft material, like a carpet/rug or a curtain, the dust falls out, making the object dust free to a large extent. This is mostly done in open air so that other things do not get dusty.
- *Sweeping*: When a broom or a brush is used to carry the dust laterally along the room, the process is known as sweeping. While sweeping any vertical surface as walls, you should remember to start from the top and sweep downwards. Similarly for lateral sweeping as for floors, start from one end of the room and move to another, preferably a door, and carry the dust all along or collect in a dust pan. All the movable objects kept on the floor should be lifted, swept under, and kept back in place.
- *Mopping*: You have read that wiping with a dry cloth is dusting, similarly, wiping a surface with a damp cloth is called 'mopping'. The piece of cloth used is known as a 'mop' and is generally coarser than a duster. In this process, both the dust, as well as easily removable dirt, is also removed. Mopping is mostly done on floors. Extra attention should be paid to nooks and corners otherwise it gets tougher to remove fixed grime later on.
- *Washing*: Sometimes mopping alone is not sufficient to remove dirt. Such surfaces are then scrubbed with the help of a yard broom along with plenty of water. Eventually the dirt loosens and is carried off by water. This process is known as 'washing'. In case of tougher stains or dirt, detergent may be added to the water.
- *Polishing*: When some reagent is rubbed on a surface to bring out the shine, the process is known as polishing and the reagent applied is known as the 'polish'. Similarly, many other objects/ decorative items made of brass, wood, marble etc, may be polished.

CLEANING EQUIPMENTS AND MATERIALS

Let us now learn about the equipment and other materials, which assist us in the process of cleaning. Can you name a few? Let us try to make a list of such equipments.

Equipments

Following are some of the equipments which you will come across during the process of cleaning:

- Brooms- brooms are either soft or hard. The soft ones are used to sweep the floors, whereas the hard ones are used to wash the floors.
- Brushes-are available in various sizes and shapes and are made of different materials. Different brushes are used for specific jobs. Brushes with nylon or plastic bristles are used for cleaning carpets or furniture, round feather brushes are used to remove cobwebs, metal brushes are used to clean wire mesh in the windows. You have special nylon brushes to clean the toilets.
- Buckets or basins- metal or plastic buckets/ basins of suitable sizes are used to carry water, detergents and chemicals so that there are no spills.
- Dust bins-these are available in plastic with a lid. These should be lined with paper so that the garbage does not stick to the surface. They should be emptied and washed daily.
- Dust pans- these are made of either plastic or metal and have flat surfaces, rounded at the sides. After sweeping, dirt and dust is collected directly into these with the help of a broom and carried to a dustbin. Dustpans save sweeping the entire amount of dust from one room to another. Instead, dust can be collected from each room and disposed of simultaneously. Dust pans should be cleaned after use.
- Dusters - These are mostly made of soft cotton, flannel or artificial feathers mounted on a stick. These are used to clean loose dust and are also used for wiping various surfaces. You should use separate dusters for dusting and wiping surfaces such as dining table, mirrors, kitchen slabs, etc. They should be washed and dried after use.
- Mops- are mostly made of thick, loosely woven cotton cloth. These are used to wipe dust from the floors. These are dipped in clean water and squeezed before wiping the floors. You should change the water after mopping each room or when it gets dirty. You should thoroughly wash the mop and spread it for drying, after use.
- Polishing cloth- these are made of soft absorbent cloth such as flannel. Dry polishing cloth helps to clean and shine the polished surfaces by rubbing them vigorously.
- Vacuum cleaner- it works on electricity and has a fan. This sucks in the dirt and dust from the surfaces and stores it in a disposable bag inside. This bag should be emptied regularly.

Materials

There are many materials and reagents, which help in cleaning, scrubbing and polishing surfaces.

Some of these are commercial preparations for cleaning and you may be already familiar with some of them:

- Water- Water is the simplest cleaning reagent available to us. Some dirt may be loosened and dissolved in it. Although most of the time, some other cleaning agent is also used along with it.
- Detergents- Detergents are available in powder, solid and liquid form.These are used with water to clean various surfaces. The basic ingredients in a detergent are surface active agents, known as surfactants. A detergent may have more ingredients to make it more effective, like alkaline salts, bleaches, foam boosters, germicides and perfumes. The exact nature and use of a detergent will actually vary just as to its ingredients. However, there are a few points which should be kept in mind while choosing a detergent. It should –
  - Be readily soluble in water
  - Be effective in all types of water and produce no scum
  - Have good wetting powers so that the solution penetrates between the object and the dirt particles
  - Have good suspending powers to suspend dislocated dirt and not allow it to settle back
  - Be effective over a wide range of temperatures
  - Be harmless to the object and the skin.
  - Clean quickly
  - Be easily rinsed away
- Abrasives- some of the common abrasives are sand, finely powdered brick, saw dust, wheat bran, emery paper, fine ash, filtered chalk etc. Besides these, steel wool, nylon mesh, coconut fibres are also used to scrub dirt. Their use depends on the surface to be cleaned and the type of dirt to be removed. The extent of cleaning will depend upon the nature of the abrasive used and on the scrubbing action.
- Acids- strong acids are used to clean toilets and are available in crystals or liquid form. Milder forms of acids are also used to clean very dirty tiles. Acids should be rinsed off as soon as possible after use and should be stored away from children. Vinegar and lemon are used to clean stains on metals like brass and copper.
- Alkalis- baking soda and ammonia are used as grease emulsifiers and stain removing agents.
- Bleaches- stains on fabrics are removed by bleaches such as sodium hypochlorite, sodium perborate, hydrogen peroxide, sodium hydrosulphite etc.
- Solvents – solvents such as methylated spirit, carbon tetrachloride, kerosene, petrol etc; are used to remove grease, wax and other stains from the surfaces. You should keep methylated spirit, kerosene,

petrol, away from fire as they are inflammable. Carbon tetrachloride is harmful if inhaled.

- Polishes- polishes are used on surfaces such as floors, furniture, leather and even metals. When rubbed on a surface, they provide a protective covering to the surface and produce shine. The object also gets cleaned in the process.

ESTABLISHING THE COLLECTIVE MODEL

Interest in employment, now the preserve of many individuals, but not all, in western economies, derives from the fact that'Work dominates the lives of men and women ... the management of employees both individually and collectively remains a central feature of organisational life'. Before beginning an examination of employment relations in the HI, it is useful to outline the historical development of employment relations and their relevance as a field of study, thereby introducing the reader to some of the key terms used throughout the book.

Hyman academic interest in employment relations was prompted when the potential stability of social order was put under threat by militant behaviour among a growing number of unionized industrial manual workers, who were no longer prepared to tolerate very bad terms and conditions of employment. This challenge to social order, which began in the late nineteenth century, was met by two responses. First, the social welfare reformers, in keeping with their predecessors who had successfully campaigned for health and safety legislation earlier in the nineteenth century, urged legal intervention to improve the conditions under which work was performed and the terms under which it was undertaken. They achieved limited success, notably the introduction of minimum wages in four manufacturing industries in 1906. The second and main response, which was to characterize public policy on employment relations until 1979, was that voluntary collective bargaining provided the best means to secure order within employer-employee relations. Collective bargaining is a process whereby employers and trade unions negotiate the substantive terms and conditions of employment, such as pay and hours of work, and procedural agreements that facilitate the resolution of disputes between the parties.

Industrial relations, the term in usage at the time, focused on the institutions of collective bargaining in fixing these'rules' of employment, largely within male-dominated manufacturing environments. Collective agreements were not legally enforceable. While public services such as the health service, the railways and the coal mines came to assume importance in industrial relations following the mass nationalization programme after the Second World War, private services remained the'Cinderella' of British industrial relations. Even though the growth of private services such as retailing and hospitality opened up more employment opportunities for women, whose main work opportunities had been in domestic service in the earlier part of the century, unregulated, female service work was deemed not to be part of industrial relations.

Even so, the lack of collective bargaining arrangements prompted the Labour government to extend the scope of minimum wage legislation to embrace these sectors. Thus in 1945 the newly named wages councils, a form of'state-sponsored' collective bargaining, were able to fix remuneration and paid holidays for many'unprotected' workers in private services.

COLLECTIVE CONSENSUS AND A MORE ACTIVE STATE

Greater state intervention in employment matters was a response by both Conservative and Labour governments to the mounting economic difficulties of the 1960s, *e.g.* statutory and voluntary incomes policies. State intervention also constituted a response to the perceived failure of voluntary collective bargaining to provide an effective regulatory mechanism for social order and social welfare, notably to protect the interests of the low-paid, many of whom were women. This perceived breakdown prompted the government to appoint a Royal Commission in 1965, the Donovan Commission, to investigate the state of employer-worker relations, in order to recommend how the'system' could be reformed.

The Donovan prescription sought to maintain voluntarism, and placed the onus on employers to improve the rules of employment, and to introduce more formal procedures for the resolution of disputes. Donovan's prescription was not universal, because it could not be applied to large parts of private services comprising small, informally managed, non-union workplaces, where female and part-time employment was concentrated. A different approach based on legal intervention in employment relations began to develop, based on employment protection for individual employees. Early employment protection rights of the 1960s included the right to a written statement of terms and conditions of employment, statutory redundancy pay and equal pay. Workers lacking the protection of a trade union and with no recourse to formal workplace procedures could resolve an employment dispute, which is those in scope of the law, by going to an industrial tribunal.

The 1970s represented a significant turning point for legal intervention in employment relations. Britain joined the European Economic Community in 1972. This heralded the start of a wide-ranging programme designed to establish a floor of new rights relating to matters including unfair dismissal, maternity leave, sex and race discrimination and health and safety at work. The main beneficiaries were to be those working in private services. Events of the 1980s and early 1990s effectively killed the model of voluntary collective bargaining. In pursuit of an overriding objective to deregulate the labour market and employment, successive Conservative governments systematically dismantled institutions deemed to interfere with the free working of the labour market, notably the trade unions and wages councils. Paradoxically, in spite of the government's antipathy to the EU's social action programme and subsequent opt out of the social chapter, the EU continued to influence British employment relations in a significant way. Rulings from the European Court of Justice obliged Britain to introduce new legislation, *e.g.* the transfer of undertakings or the amendment of existing legislation relating to equal pay and sex discrimination.

The floor of employment rights was both strengthened and extended. Managers reasserted the right to manage increasingly flexible and non-standard workers under the banner of'managerialism', in workplaces that might be labelled'bleak houses'. An alternative version of management thinking stressed the benefits of'commitment' over'control'. Both approaches came to signify the two variants of HRM.'Soft' HRM emphasized fostering commitment, improving quality and developing the human resource, whereas'hard' HRM was contingent and calculating in its utilization of the human resource.

If organizations were to survive the effects of adverse economic conditions, globalization and increasing competition, the imperative was to integrate HRM within business strategy. The impact of HRM on industrial relations was widely debated. Other key issues in the wider academic debate included the extent of continuity and change in industrial relations, the sharp decline in trade union membership, the impact of deregulation and whether employment relations could be re-regulated.

NEW LABOUR: NEW HOPE?

By the mid-1990s individual relationships were catapulted firmly to the forefront of analysis of the employment relationship. Recognition of this change had been apparent from WIRS in 1990, perhaps most notably within the HI. HI managers are free to exercise a high degree of managerial prerogative in the absence of unorganized labour, termed'unbridled individualism'. The election of a Labour government for the first time in nearly 20 years in 1997 raised expectations that there would be a new agenda for employment relations, although Heery's assessment was that'it is extremely doubtful whether New Labour will issue in a new industrial relations'. New Labour's stakeholder economy is based on fairness and partnership. Fairness at work is to be achieved in two ways.

The government signed up to the EU social chapter and set about introducing a new floor of minimum employment standards, including a National Minimum Wage, and family-friendly measures. Social partnership between employers and workers is designed to foster a more consensual and cooperative relationship between employers and employees. The Low Pay Commission whose first task was to recommend the initial rate of the NMW, provides an early manifestation of social partnership comprising employer, worker and independent representatives. Although many of the Conservatives' trade union reforms remain in place, the introduction of statutory trade union recognition procedures might help reverse the steep decline in trade union membership.

By the time of WERS in 1998 the system of collective representation had crumbled'to such an extent that it no longer represented the dominant model'. In reality employment relations could conform to different and diverse patterns. Private service establishments employing 25 or more employees were numerically more important than private sector manufacturing and the public sector put together. Their share of employment increased from 26 per cent in 1980 to 44 per cent in 1998, reinforcing the point that alternative ways to view and reform employment relations were long overdue, particularly in circumstances of'bleak house' or'black hole' employment. Although we find these terms wanting in respect of the HI, they highlight the relevance of the industry as a unit of analysis. Consequently we shall show how these types of workplaces throw up major problems for employment relations reform.

Agenda for the Twenty-first Century

In calling for a new industrial relations paradigm, Ackers now argues that the new problem of social order focuses on links between employment and society, and that such a link provides an explicit ethical framework for policies like social partnership. He rejects the traditional industrial relations notion of workers as unattached individuals in their out-of-work lives, and argues that industrial relations can no longer ignore issues of work-life balance and corporate social responsibility. Indeed social concerns underpin'Fairness at Work' and the'Welfare to Work' programme, and family-friendly issues are a new addition to WERS.

Hence a new definition of industrial relations as neo-pluralism: Employment relations are the study of the social institutions involved in the normative regulation of the employment relationship and business's interaction with other stakeholders in society. Thus Ackers rejects as inappropriate Kelly's industrial relations paradigm for the twenty-first century, which derives from a redefinition of Marxism based on socialism, workers' mobilization, economic militancy and strikes, and organized labour. Edwards identifies three pressing issues in contemporary employment relations:'high commitment' or'high involvement' work systems, the international context and economic performance.

The first, although interesting, is very rarely found anywhere in Britain. Its alternative of'low skills' and'low wages' strikes right at the heart of much hospitality employment. This links to economic performance, where the absence of collective bargaining is likely to have contributed to income inequality and the perpetuation of low pay in the HI, although pay may be subsidized by the state through social security and taxation. The further subsidy of low pay through tips as a defensible employment practice is a matter of conjecture. We shall also explore if particular employment relations practices can be linked to successful performance outcomes. The international context and its implications for employment relations in the HI are considered below and in subsequent chapters.

A fourth pressing issue can be added. Employment relations discourse needs to recognize that prejudice and bias have been built into much of the theoretical and practical analysis, thus distorting its perspective. Gender is not the only example, but may be the most obvious. In spite of an increasing interest in what may be described as'women's issues', such as (un)equal pay and employment opportunities, family-friendly policies and sexual harassment, one major barrier to understanding employment relations is an assumption that they are gender neutral. The argument is that adding women's issues to the agenda is simply not good enough. Management, trade unions and the state are not gender neutral, and therefore we need to recognize the gendered characteristics of the employment relationship and work and integrate this into our understanding of the field of employment relations. Other'omissions' include age, ethnicity and the role of customers.

We shall explore these issues throughout the book where it is possible or relevant to do so. All these issues were placed under review in The Future of Work Programme launched by the ESRC in 1998. The Programme has supported 27 projects designed to rectify gaps in our understanding and improve the quality of information available to the policy-makers in the UK. Topics under investigation have included the future of unskilled work, business re-engineering and performance, the changing position of ethnic minorities and women in the labour market, the future for trade unions and the changing nature of the employment relationship.

EMPLOYMENT RELATIONS IN THE HOSPITALITY INDUSTRY

Three terms denote the relations between managers and workers in the employment relationship-industrial relations, employee relations and employment relations. These terms are often used interchangeably, but can also convey subtle differences of meaning.

They may coincide with other fields of academic enquiry and practical activity concerned with'people management', namely personnel management and HRM. Edwards provides an insightful analysis of the employment relationship, taking as his starting point the distinction made by Fox and Flanders between market relations and managerial relations. At the root is an economic exchange between capital and labour, in which the price of labour is set as a contract of employment. In this economic exchange between the buyer and seller of labour, the parties do not share equal power resources. In common law the employer has the right to command and the employee has a duty to obey. The commodity at the heart of the bargain is the worker's labour power. The employer will seek to maximize control over that'labour process' in order to generate a surplus as profit. The employment relationship, as an exchange and in recognition of its broader context, has also been termed the effort-reward bargain:'an economic, social and political relationship, for which employees provide manual and mental labour in return for rewards allotted by employers'. Labour only becomes useful if it can be persuaded by management to work, but this is only the beginning. Workers must demonstrate commitment, continue working to the required standards, and not deviate from those standards.

In other words workers must follow'rules', otherwise management may need to deploy corrective or punitive measures via the disciplinary procedure. Bonamy and May argue that a weakening of employment relationships since the 1970s has given rise to the emergence of employment as a service relationship. This relationship demands increased recognition of the professional qualities of the'autonomous' worker, which poses problems of incompatibility with an employment contract built upon subordination. Pay is determined by time worked, whilst idle time due to poor organization and absenteeism is reduced. This is manifested in new forms of employment contract, externalization of employment to agencies and the sub-contracting of activities. Edwards notes that, if we were starting from scratch, 'employment relations' might be the best label.

Employment relations do not rule out all variants within the employment relationship including:

- Trade unions and formal collective bargaining;
- Individually based management/workforce relations conducted informally;
- Managerialism;
- More democratic and highly participative non-union relations;
- Men, women and disadvantaged groups;
- Employees and workers, including atypical workers and the self-employed.

Further:

- Employment relations is the main term used in the WERS sourcebook;
- Industrial tribunals have been renamed employment tribunals;
- The cornerstone of New Labour's industrial relations policy is the Employment Relations Act 1999.

A necessary departure for this book, as noted earlier, is to relocate the nexus of the employment relationship to include a relatively ignored third actor in the employment relationship-the customer. The notion of the customer in the employment relationship has been increasingly incorporated into the sociology of work, but less so in employment relations. Front-line workers, such as receptionists and servers in bars and restaurants, have to serve two'masters': their superior manager and the customer. Individual workers can have a simultaneous and coterminous employment relationship with the organization and the customer.

Organizations in services are best seen as inverted pyramids, with most workers in direct customer contact. Direct service workers engage directly with customers in an exchange that carries both economic and social connotations. Their ability to deliver successfully hinges upon a social relationship with indirect service workers, whose actions are also instrumental to the provision of good customer service, *e.g.* an enjoyable meal or clean bedrooms. Indirect service workers are not in regular customer contact, so customer influence may be more economic than social. Hence, customers cannot be excluded from an analysis of the employment relationship.

An earlier definition has been revised:

- Employee relations in hotels and catering are about the management of employment and work relationships between managers and workers and, sometimes, customers; it also covers contemporary employment and work practices.
- Before exploring the facets of the employment relationship, we need to outline why our attitude towards things influences the way in which we see any given situation, and how it.
- Triplets watch a local football match from adjoining seats, getting an almost identical view of the game. The result is United 5 City 1. One triplet is deliriously happy, the second feels very low, while the third is able to provide a balanced analysis of events, conceding that the result was a fair one, although two of United's goals were the result of dubious refereeing and City deserved more than a single goal.
- Why did their particular attitude affect their view of the game?

Fox proposed two frames of reference as a means by which'the problems of industrial relations can be seen realistically and laid more open to solution'. The unitary perspective is a'management ideology' built on the belief that everyone in the organization shares the same goals, and that'conflict' is pathological and derives from deviance. Trade unions are seen as an intrusion, competing with management for worker loyalty.

Fox's main argument was that the unitary perspective was a naive and unrealistic frame of reference that might'distort reality and thereby prejudice solutions'. Yet in reality many managers do perceive their organizations in unitary terms, regarding themselves as the sole source of authority. Unitarism has underpinned the'human relations school' of management, including Mayo, Likert, McGregor, Schein and Herzberg, and reasserted itself in'managerialism' and HRM. Fox suggested that the more realistic approach to managing people was to recognize that organizations are pluralistic, comprising various groups, each with their own basis of authority and sets of interests.

The'rules' of employment are not just the preserve of management. A new pay rate set by management will not necessarily be seen as fair by workers, creating an issue of potential dispute. Therefore conflict or differences between individuals and groups are inevitable. Management should recognize this inevitability, and find the ways and means to regulate such differences. An institutional approach-collective bargaining between employers and trade unions, and the development of formal procedures to deal with disputes about pay, grievances and discipline-was considered to be the most appropriate solution. However, this is flawed to the extent that it implies both parties to the bargain have equal power resources at their disposal.

Later Fox revised his thinking and added a third perspective of radicalism, prompted by a wave of'shop floor' discontent and'wildcat' strikes at workplace level. Such worker behaviour was perceived as a reaction against exploitative and oppressive employers whose sole aim was to maximize profit. Conflict was caused by the economic disparity of society as a whole, with the principal disparity between capital and labour-employers who own and manage the means of production and workers who have their capital to sell.

This view underpins the labour process approach. This approach stresses the contradiction of managerial goals, with regulation and control having to be balanced by the need to gain workers' consent. Even today Edwards argues that unitarism cannot be written off as naive and outdated any more than radicalism because of the apparent disappearance of discontent. Ackers' neo-pluralism refocuses the employment relationship beyond the workplace by connecting the old pluralist and voluntary frames of reference with new questions raised by contemporary society. The health of society is put first, encouraging industrial relations policy initiatives that are driven by social concerns, not just a business agenda. Further he argues that the employment relationship bears hidden ethical. considerations of trust and responsibility in relation to human beings.

As we shall show, both managers and workers in a variety of work and employment situations in the HI do see their workplaces in unitary terms, but this does not necessarily infer the absence of conflict. Areas of potential dispute, conflict and difference do exist between managers and workers, between managers and other managers, and between workers and customers, demonstrating that workplaces are pluralistic. In cases where workers'fiddle' or'pilfer' from their employer, the nature of their behaviour is more in keeping with a radical perspective.

Thus it is possible to observe facets of unitarism, pluralism and radicalism in the same employment relationship in which management, for the most part, remains the more powerful. Even so, areas of common interest self-evidently exist otherwise all these relationships would break down. Consent provides the basis for resolving conflict and achieving cooperation. Cooperation is built on trust between individuals engendered at workplace level rather than through elaborate organizational mechanisms. Yet securing workers' consent is neither a straightforward nor certain process. Therefore, a mix of overt and covert conflict and cooperation underpins all employment relationships and, as we shall argue, workplace harmony owes more to pragmatic acceptance and accommodation among the parties in the employment relationship than to ideological belief.

THE RULES OF EMPLOYMENT AND POWER RELATIONS

We have already noted that the employment relationship is underpinned by rules, hence the continuing validity of Clegg's definition of industrial relations as'The study of the rules governing employment' which Edwards explains in more detail: This does not limit the subject to the collective relations between managements and trade unions, for a rule can derive from other sources, and there are rules governing non-union groups; nor does it restrict analysis to one sector, for it covers all paid forms of employment. A rule is a social institution involving two or more parties which may have its basis in law, a written collective agreement, an unwritten agreement, a unilateral decree or merely an understanding that has the force of custom. In non-union settings, as much as union ones, rules determine rates of pay, hours of work, job descriptions and many other aspects of employment. The subject is about the ways in which the employment relationship is regulated. To regulate means to control, to adapt or adjust continuously or to adjust by rule.

MANAGERIAL ISSUES

While rules may be the substantive rules of employment, *e.g.* pay and conditions of employment, implicit in the notion of rules affecting people is the concept of behaviour. Management's job is to control and direct workers' behaviour to perform work to the desired standards, and thereby ensure that the rules of employment are adhered to.

Four key issues arise:

1. Rules are not always absolute and may be gendered.
2. Managerial control of workers' behaviour is underpinned by a power relationship.
3. This power relationship is unequal and may be gendered.
4. Managers have a choice of means to maximize control.

The first point is that one should caution against perceiving rules in too absolute a sense. At one end of the spectrum rules embodied in the law of the land provide a good example of formal rules. Any breach may incur very severe penalties, *e.g.* health and safety. In a workplace setting rules in practice may derive from informal understandings that can in one set of circumstances be interpreted by the worker as a permissive concession or in a different set of circumstances as something to be observed at all costs. Strawberries as a worker's perquisite during the Wimbledon lawn tennis championship are a good example. Experienced workers know that taking home unwanted strawberries is'permitted' during busy periods. When fewer staff are needed, increased managerial surveillance will be deployed to dismiss staff caught in possession of company property.

Rule-learning is part of what Polanyi refers to as'tacit skills'. As argued elsewhere:

Tacit skills, such as learning to deal with customers, are learnt in and through the very act of doing, often involving trial and error and not from following a body of procedurally-designed rules. They are seen as an interpretive achievement of the user as to how the'rules' fit the task in hand. By mastery of the rules comes the power to extend them. This example embracing the customer provides a developmental point to Edwards' observation that rule-making is difficult, and that rules have to be interpreted in action for them to have any real meaning. Specifically in the labour contract this is because the worker's ability to work is only realised as useful labour in the course of carrying out that work, hence'a rule is a complex social institution'.

In service work physical appearance and'personality', or'aesthetic labour', are an implicit part of the employment contract. Only female flight attendants, not their male colleagues, are subjected to regular weigh-ins to ensure they comply to specified weight: height ratios. This demonstrates clearly how a rule may be gendered. The second point to note is that the very essence of management seeking to control workers' behaviour is underpinned by a power relationship. Power is the capacity to pursue one's own interests individually and collectively, involving the capacity to oppose the actions of others and to pursue one's own objectives, and is embedded in continuing relationships. This does not mean power has to be exercised by either party in an overt sense. The threat of power may be sufficient to maintain broadly consensual employment relationships, such that any disputes or differences are resolved amicably without recourse to either party seeking to deploy sanctions against the other. The third point assumes a power inequality in the employment relationship. Self-evidently an employer is more powerful than an individual worker. The employer's ability to terminate a worker's services is likely to be more detrimental to the worker than to the employer, in spite of employment protection legislation.

Yet the individual behaviours of workers, such as high labour turnover, may be detrimental to an employer, even though they are not concerted. When workers combine collectively, with or without the backing of a trade union, there is some tilt in the balance of power, because collective sanctions may be imposed against the employer. Ultimately the outcome of the process by which each side seeks to gain concessions will depend on the relative power of the parties. For example, a plentiful supply of suitable workers in the labour market makes existing workers more readily dispensable and replaceable on the employer's terms. The opposite would be true for workers with scarce skills who can command high wages.

As Wajcman argues, gender relations are power-based and women's subordination in the workforce and workplace owes as much to trade unions as it does to managers. Spradley and Mann provide a graphic account of how the subordination of one group of female workers was brought about by another group of male workers who were the custodians of the male proprietor's trust. The male bartenders controlled the orders, and sought to make the cocktail waitresses' job difficult by giving orders in an inconsistent and confusing way. Any mistakes became the waitresses' responsibility, even if they had been caused by the bartenders.

Such was the power of the bartenders that pleasing them became more important than pleasing the customers. The fourth point is that managers have a choice of means to maximize control over workers. Friedman's'direct control' is a variant of Taylorism. Management is responsible for planning, designing and organizing the labour process, while cheap, unskilled workers perform standardized, simple repetitive tasks. Fast food is a good case in point, and also epitomizes McDonaldization, a social critique of how contemporary society and culture are being shaped by rationalist scientific management. While Taylorism sought to control the organization of work, McDonaldization is based on rationalization, replication, standardization of products and service, and quantification. In this low trust strategy worker behaviour is controlled through the use of standardized scripts in the service encounter.

In Friedman's alternative of'responsible autonomy', a high trust approach, managers delegate control to relatively privileged skilled workers who may already have elements of job control and discretion. The objective is to get workers to identify with the competitive aims of the organization so they will behave responsibly with minimum supervision. An obvious example of where such an approach might be used is in a luxury hotel, but it is also associated with empowerment and much customer-service work. As we shall argue and implied in the example of cocktail waitresses, these and other similar approaches including'hard' and'soft' HRM provide a useful framework for analysis, but are not necessarily alternatives.

The history of hotel internationalization has been characterized by American chains that secure control and integration through highly standardized procedures and manuals of operational procedures. Yet a'soft' focus on the service encounter as the driver of competitive advantage necessitates developing a culture of customised service. Mass customization illustrated by

Burger King's have it your way' slogan as a challenge to McDonald's hold on the market is proposed as an alternative paradigm to McDonaldization.

WORKERS AND CUSTOMERS

Other tensions within the employment relationship impinge upon the rules of employment and power relationships. If management is about the achievement of organizational goals through people it can be argued that managers will be successful to the extent that these goals coincide with the aims and aspirations of those people, be they workers or customers. This'matching' of broadly reciprocal needs between employers and workers may be referred to as a'psychological contract', or set of contracts. It suggests managers and workers can share goals, but this is not at all straightforward. There is not a necessarily clearcut distinction between boss and worker, or a'them and us' scenario. Further we must also account for a psychological contract with customers.

Two key points are noteworthy:

1. Organizations comprise people and are, therefore, social organizations.
2. People, as social animals, may behave in unpredictable ways.

Workers

Human beings do not necessarily behave consistently or predictably, even in the same sets of circumstances. People are citizens and customers as well as employees, and these multiple identities bring different and sometimes conflicting expectations of the organization.

This makes the management of the employment relationship an uncertain process within which there is a blend of contradictory principles around the need to control and to gain the consent of workers. Workers may seek to regain control individually or collectively when they perceive that management has operated outside the rules.

At that point workers' consent has been withdrawn and management will need to find ways to restore order and regain consent. In Lucas workers' individual response to organizational rules is seen in three main ways-to conform or be deviant in employment, or to terminate their employment. These responses are similar to Marchington's'getting on','getting by' and'getting back'. These are behaviours deployed in circumstances where customer care and service quality are dependent on workers' use of their tacit skills, which contain both technical and attitudinal elements. Limiting the definition of tacit skills to employer-employee relations is too narrow. Marchington overlooked how workers exhibit their tacit skills in ways other than in respect of their relationship with the employer, notably the customer. The point that'getting back','getting by' and'getting on' are as much resistance strategies in the labour process as coping mechanisms is developed.

At workplace level personal relationships are likely to be closely connected to morale and success. Managers often'muck in' when required. In small workplaces the existence of a single leader, often the owner, may serve to inspire loyalty from the workforce, but it is not a one-way process, as workers' respect has to be earned. Is it realistic to suggest that Mina, Jo and Sadie, who wait on table in the restaurant, share all the same goals as their boss? The hotel may not be doing very well, so there may be mutual concern for the survival of the business. Yet these ladies' main goal may be to serve their customers cheerfully and effectively, while at the same time enjoying some social banter among themselves and with their customers in the process of earning a reasonable wage.

Customers

Within the triadic employment relationship a simultaneous and coterminous relationship with the organization and the customer directly impinges on how workers carry out their work, and such interactions may be rewarding or stressful. The consequent effect on workers' performance may have positive or negative implications for the rules of employment: what they can earn, their prospects of promotion or actually keeping their job. Unequivocally the worker-customer relationship affects the rules governing employment and workplace behaviour. But so do employer-customer relationships, hence the employment relationship embodies a triadic set of power relations. This relationship embodies a socio-economic exchange, and is not simply an economic exchange around the price of labour.

Fox provides a useful starting point, since he noted that organizations are social organizations and how people behave is a crucial issue in the employment relationship. Even Edwards' point that'a rule is a complex social institution' does not adequately encapsulate our position. The main justification for widening the scope of this relationship derives from the fact that the service encounter is the interaction of the producer and consumer of services, and is a more complex phenomenon where financial considerations are interwoven with social ones. In hospitality the social function of service work derives from the provision of a'home away from home'.

The service encounter entails'emotion work'-the assumption of a social-self, which effectively masks the individual's own personal dispositions to act, including the need to smile and be pleasant in an uninvolved way. We have already noted that'aesthetic' and sexual labour may also be inherent in service work. It is the'normalizing' social role of service labour that distinguishes it from other wage labour. Service work cannot be understood in terms of economic rationality alone.

Examination must be based on the supposition that service work is the intended outcome of a necessarily social process in which some social interaction occurs between one or more producers and one or more consumers. The relations between three groups of people-managers, workers and customers-embody the potential for contradiction between, on the one hand, uncertainty, unpredictability, conflict and difference and, on the other hand, consent, team effort and concerted performance. The practical benefit this book seeks to convey accrues from an understanding of the nature and scope of the rules of employment in this triadic employment relationship, and how it is regulated, primarily at workplace level.

THE EMPLOYMENT RELATIONSHIP IN A WIDER CONTEXT

This chapter concludes by considering some key external contextual influences on workplace employment relationships at two levels-internationally and, in more detail, nationally in Britain.

The International Context

The national context of British employment relations increasingly needs to be understood within a much wider international context.

Three international dimensions have particular resonance for this book:

- International competition has created more open economies that have attracted investment from foreign-owned businesses. For example the French-owned groups Accor and Envergure have respectively opened hotels within their Novotel and Campanile brands in the United Kingdom.
- On a larger scale American multinational corporations have created world brands. McDonald's, Burger King, KFC and Marriott are among those that are now household names in many countries across the world.
- Spin-offs from European integration, especially on employment law in Britain, have provided an important underpinning to the employment relationship in the HI.

Foreign investment and MNCs are clearly important factors underpinning the expansion of hospitality and tourism not only in Britain but also in developing countries. Examples ofbetter' employment practices, in so far as they may exist in the British HI, have been associated with foreign-owned businesses. Aspects of the American model of employment relations, that is non-union and market-driven, may seem to reflect some aspects of observed employment relations practice in the British HI, but the similarity has been overstated.

The United States has substantially more legal regulation than Britain, which has benefited American HI workers, while the trade unions are not entirely powerless-issues we highlight in later chapters. The European model based on social partnership designed to forge a common agenda between capital and labour is considerably more diverse and different across the member states than is often acknowledged. While we cannot expect it to reflect current developments in HI employment relations in most British workplaces, it has not necessarily produced wholesale benefits for HI workers across the EU either.

The British experience is not necessarily mirrored in other countries across the world. Differences in other countries' institutional arrangements and cultural considerations are among the factors that will affect their employment relations systems. Detailed comparison with other countries is beyond the scope of this book, but key instances of international employment relations within hospitality and tourism are cited throughout the remaining chapters.

THE BRITISH CONTEXT

Workplace employment relationships cannot be immune from wider economic, social, legal and political contextual influences. Contemporary examples, which may be influential in Britain today. The distinctions between these sets of influences are not always clear-cut as they can be interrelated.

The State

Although we have already touched upon some aspects of the state's interest in employment relations, we need to examine its role in a little more detail. The state is not a single or cohesive body, and comprises a number of institutions that have an interest in the employment relationship, whose objectives do not necessarily coincide. Parliament is the legislature, government ministers form the executive, the judiciary enforces the law, and civil servants are the administrators. The state sponsors specialist agencies in the field of employment, and has done so since the end of the nineteenth century. Three government departments impinge on employment relations within a much wider brief. The most important is the DTI, which has overall responsibility for employment relations, small firms and competitiveness. The Employment Relations directorate is responsible for developing policy and legislation affecting individual workers and trade unions, EU legislation, promoting partnership and best practice, regulation of employment tribunals and the dates of public holidays. The DTI publishes consultation documents, research papers, practical guidance on how to implement employment legislation and regulations, and codes of practice on matters such as picketing.

The Department for Education and Skills is responsible for developing the skills of young people and adults. he Department for Work and Pensions delivers support and advice in areas of work and work-related benefits, including New Deal, sickness and accidents at work, and retirement. Although publicly funded, other state agencies and bodies are independent of government because they are controlled and managed by their own executive. The main institutions discussed later in the book are ACAS, the Central Arbitration Committee and employment tribunals. The Equal Opportunities Commission, Commission for Racial Equality and Disability Rights Commission each have overall responsibility for specific types of anti-discrimination or equal opportunities legislation.

The Health and Safety Commission and Health and Safety Executive have responsibility for health, safety and welfare legislation. Their roles include the publication of codes of practice. Other bodies assist with the enforcement of minimum employment standards. The Inland Revenue's powers include obtaining information from employers, issuing enforcement notices requiring employers to pay the NMW and imposing penalties on employers not observing the NMW. Environmental Health Officers are responsible for the enforcement of health and safety standards. Many other institutions, some of which may have a political bias, offer a mixture of fact and opinion on employment relations. National bodies, which take either an employer or management view, include the Confederation of British Industry, the Institute of Directors, and the Chartered Institute of Personnel and Development. The British Hospitality Association, Restaurant Association, the British Beer and Pub Association the Hotel, Catering and International Management Association and the British Institute of Innkeeping are specific to the HI. The HtF, formerly the HI's National Training Organisation is recognized by government as the employer-led voice on all issues relating to hospitality training, education and qualifications.

The HtF also carries out research, and produces useful statistical information about the labour market. The Trades Union Congress Institute of Employment Rights and the Low Pay Network serve to defend workers' interests. The HCIMA can also be regarded as having a worker perspective since, as the professional body of hospitality managers; it serves to defend their interests as well as disseminating good management practice. Three large unions have special sections for hospitality workers: the General, Municipal and Boilermakers' Union, the Transport and General Workers Union and the Union of Shop, Distributive and Allied Workers. The National Association of Licensed House Managers was self-standing for many years, but has recently become part of the TGWU.

SUPPLEMENTARY HOUSEKEEPING KNOWLEDGE

Key is a specially shaped and designed piece of metal or plastic/magnetic card used for opening and closing the lock.it is an essential too for lock. In a hotel, it refers to the door lock key, which is primarily concerned for safety of hotel room contents and guest valuables.

The responsibility for handling key of the houseguest has always been with front office and housekeeping department. In all large, hotels the information desk holds the responsibility of issuing, retrieving and controlling room keys.

Room attendants also handle room keys to prepare guest bedrooms. Hence, it is essential that strict control of keys is maintained regularly.

MANUAL/MECHANICAL KEY SYSTEM

The standard of manual key system in operation comprises of the following:

- Individual guestroom keys.
- Sub master key/floor key
- Master key
- Grand master key
- Emergency key
- Electronic/Magnetic Key System

CONTROL OF ROOM KEYS

Maintaining a strict control over the guest room keys is the responsibility of the information section. it is very necessary that a strict control be maintained for averting any mishaps. The information assistant keeps the keys into the correct slots of the key rack/pigeon hole.

For the security of both guest and the hotel, room keys issuing system must be very efficiently operated and controlled.the theft, loss or unauthorized duplication of keys could have serious security implication.

The system of key control is as follows:

- Know the different types of keys in circulation within the hotel.
- Keys should not be handed over to unauthorized personnel.
- Guests should be requested to handle their key cautiously.
- Guests should be urged to drop the keys at the'key drop slot' maintained at front desk each time they leave the premises upon their departure or for their own personal reasons.
- Identification should be sought and verified, if necessary, prior to issuing a key to a guest.
- key should always be placed in the correct hook/slot of the rack
- A regular inventory of keys should be carried out, and missing or damaged key should be reported immediately to the concerned authority.
- The master keys, when not in use, should be secured in a safe place.
- Key should be retrieved from the guests at time of check out.
- A logbook or control sheet can be used to monitor the distribution of keys.
- The keys of inter-connecting rooms should be kept secure and scheduled.

ISSUING KEY CARD

In many hotels, keys are stored in key rack located at the front desk. The rack consists of a number of'pigeon holes' large enough to hold regular size mail, with a key hook above. Each pigeonhole is clearly numbered by floor and by room in its console.

ISSUING KEYS TO INDIVIDUAL

Room keys are issued to guest on completion of the registration formalities. The room key may be issued to the guest personally, or to the staff escorting the guest to the room *e.g.* porter/bellboy. At all other times keys should only be issued to guest on presentation of identification, *i.e.* the key card.

ISSUING KEYS TO GROUP

For groups, keys may be organized in advance and distributed to its members in individual envelopes.

Table. Specimen of Key Control Sheet.

KEY CONTROL SHEET No................................						Date:	
Key code	**Name**	**Signature**	**Time out**	**Issued by**	**Time in**	**Signature**	**Received by**

LOST AND FOUND MANAGEMENT

Lost and found refers to those articles which are unintentionally left or lost by a guest in the hotel premises. Such articles can range from jewellery, costly electronic goods to simple garments etc. The hotel considers it duty to secure such articles until the guest claims them.

PROCEDURES OF HANDLING LOST AND FOUND ARTICLES

- It is general practice that any lost properties found in room/floor or the public areas should be handed over to the lost and found section of housekeeping department immediately
- The appropriate details should be entered into a slip, which is put along with the articles in a transparent bag.
- Control desk supervisor or assistant housekeeper fills up lost and found register.

- The article is sent to the lost and found section, where it is usually kept for a period of 6 months to one year or depending upon the policy of the management.
- Perishable articles or stuffs (non-valuables) are kept for a day, and then it is handed over to the finder or distributed to the concerned area staff after the set time.
- In case of jewellery (valuables) being found it is normally secured from 6 months to 1 year. If it is not claimed by the guest within the stipulated time, the item is sold and the housekeeping staff are distributing the amount.
- Incase the owner claims for the lost articles, proper identification is asked to establish the real owner. then the lost and found register is filled-up
- The signature if the guest is obtained and the article is handed over.
- The lost and found articles are normally disposed in the following ways
- The owner may come to claim or authorize someone on his. Her behalf to collect.
- It may be sent through the mail/courier as per the instruction of the owner
- If the article is not claimed within the specified time, it may be auctioned or given to the finder.

Table. Specimen of Lost and Found Slip/Form

Lost and Found Valuable/Non valuable No	
Finder's Name................................. Location /Room no............................ Description of article .... Name of the guest...................................... Address.. Signature of depositor	Date......................... Time......................... Signature of receiver
A. Received by owner Name................................ Address............................ Date.....................................	 Telephone No:........................ Signature............................
B. Dispatched by Post, Postal receipt No............. Name .... Address............................ Date................................	 Telephone No......................... Signature..............................
C. Retrieved by finder Name... Cloak No....................................	 Signature............................ Gate pass No........................
Property handed over by Name.................................. Signature............................	 Designation.......................... Date..................................

Table. Lost and Found Record.

No. Lost and Found Record								
Date / time	Received be Owner			Received by Finder			Dispatched by	
	Name	Address	Signature	Name	Signature	Gate Pass No.	Name.	Signature

PEST CONTROL

Pest conrol is a programme of eradicating pests. Pests are removed through pesticides.a pesticide is called a chemical used to control or eliminate pests. Insects are probably the larger pests.many insects transmit diseases such as malaria and typhus.some insects destroy or cause heavy damage to valuable vegetation. Other common pests include bacteria, fungi, rats and such weeds as poison ivy and ragweed.

Manufacturers use various chemicals in making pesticides:

- A pest is any organism that competes with humans, domestic animals. Or desirable plants for food or water.
- Injures humans, domestic animals, desirable plants, structures or possessions
- Transmits diseases and annoys humans or domestic animals.
- Pest control goals
- *Prevention*: Keeping a pest from becoming a problem
- *Suppression*: Reducing pest numbers or damage to an acceptable level.
- *Eradication*: Destroying an entire pest population.

BED BUGS

These are tiny parasitic creatures that feed on the blood of human and other animals. They are more of a nuisance than a danger, although they can cause severe irritation in some people, leading to loss of sleep and lack of energy particularly in children they hide in the crevices of beds, furniture, upholstery, wallpaper and skirting boards, emerging when hungry.

CONTROL MEASURES

Pouring boiling water into crevices is a temporary solution to get rid of bed bug. a thick application of kerosene of emulsion is also effective against bed bugs.

BEETLES

- *Biscuit Bittles*: these are brown with dense layer of yellow hairs on their body. They pests that attack cereal products and have even been known to infest poisonous substances such as strychnine.
- *Carpet Beetles*: these are the beetles that will attack furs, carpets and all kinds of woolen textiles. Carpet beetles are pests of animal and

occasionally plant, products. They flourish in situations when they remain undisturbed-*e.g.* beneath carpets, around skirting boards and in wardrobes.

- *Control measures*: frequent vacuum-cleaning of fluff and dust is essential for control of these pests. in case of an infestation, insecticide powders may be sprayed in the affected areas.
- *Furniture Beetles*: This beetle makes its home in crevices and cracks of furniture made of unpolished wood.
- *Control measure*: the best time to apply insecticides to wood is around March o May, when the insects are near the surface of wood just before they emerge and so are more susceptible to the insecticide.
- *Cockroaches*: these are several pieces of nocturnal insect that spend most of the day hiding in cracks around drains or in other dark secluded crevices. Two common species of the cockroach are the German and the Oriental.
- Cockroaches carry food-poisoning bacteria in their bodies and are responsible for the spread of dysentery and gastroenteritis. They will feed on almost anything, from faecal matter to food for human consumption.
- *Control measures*: cockroaches are the most difficult pests to eradicate.Proprietary cockroach-killer preparations may be used in the infested areas.however, pest-control experts need to be called if the infestation persists.
- *Fleas*: This area common nuisance. Even a vacant house can have a flea problem. They are usually laid in a host's bedding, hatching at room temperature in about 10 days.
- *Control measures*: on finding an infestation, the property should be sprayed with an insecticide that will kill adult fleas, their eggs, larvae and pupae. Pest-control expert should be called in to help with the problem.
- *Moths*: some of the moths species attack stores food products and are often found in houses, hotel kitchens, bakeries, grain stores and warehouses-anywhere that food is stored, especially dried foodstuffs such as grains, cereals, flours, nuts an so on..
- *Clothes Moths*: clothes moths in general are nocturnal insects and, although males and spent females sometimes come into the light, they are more likely to scuttle for cover that to fly into the open when disturbed.
- *Control measures*: cloth moth damage can be minimized by treating fabrics by mothproof finish.spray may be applied to carpets. Especially along and beneath the edges adjacent to skirting underneath furniture, and other likely areas of infestation where prolonged contact with

humans or pet animals is unlikely-clothing and bedding should not be sprayed with insecticides.

- *Ants*: these insects generally invade in large numbers when they come in search of food, especially sweet substances.
- *Control measures*: they can be destroyed by placing tablespoonfuls of carbon bisulphde at the entrance.
- *Flies*: these filthy are dangerous to health as the contaminate food, causing diseases such as typhoid, cholera, dysentery and so on. They carry the disease germs on their legs and in their saliva. These are transferred to food on which they sit.
- *Control measures*: to eradicate flies, the first essential step is to destroy all possible breeding grounds early in the year before egg-laying begins. All garbage should be burnt. All dustbins should be covered. And a good standard of cleanliness for the surrounding should be maintained. Aerosol fly-killer sprays are also effective.
- *Mosquitoes*: these transmit diseases such as malaria, filarial and yellow fever. As life cycle of mosquitoes begins in water, do not allow water to stagnate in and around the property. The pits should not be open.
- *Control measures:* cover drains and pour kerosene oil onto these to prevent larvae from thriving there and growing into adult mosquitoes. Fine gauze on windows prevents the entry of mosquitoes. An effective, eco-friendly method for the control of mosquitoes is to place pots of water around the property for a week to two.during this time, the mosquitoes lay their eggs in the water. Before the eggs can develop, and however, this water is discarded, killing the larvae.
- *Mice and rats*: rats and mice both carry disease germs and may cause food poisoning, infection, jaundice and so. mice and can contaminate food stuffs, work tops and utensils with their urine, droppings and fur and are responsible for spread of many diseases, some of which can be fatal to humans.
- *Control measures*: the most effective methods of controlling these pests are poisoning, trapping, fumigating to eliminate their food supply and shelter and rat-proofing buidings.proprietary poisons are available to destroy rats and mice as well.
- *Fungi*: some fungi in the form of wet or dry rot can cause considerable structural damage to a property.
- *Wet Rot*: it is more likely to be found in kitchens, bathrooms and roofs. Outsides, it attacks window frames and sills, doors and door frames where water has penetrated the paintwork.
- *Control measures*: it is best to call pest-control firms to deal with wet rot.timber can be treated with water-repellent preservations to prevent wet rot.

SAFETY AND SECURITY

Safety and security are concepts often used interchangeably, and it should be understood that both are means of safeguarding human and physical assets. The tern'safety' is with reference to such things as disasters, emergencies, fire prevention and protection, and conditions that provide for freedom from injury and prevent damage to property. The tern'security' is used with reference to freedom from fear, anxiety, and doubts concerning humans as well as protection thefts of guest, employee, or hotel property.

WORK-ENVIRONMENT SAFETY AND JOB-SAFETY ANALYSIS

The management of any place of work are legally bound to provide a hazard-free environment to their employees. The nature of work that the housekeeping staffs are involved in is such that employees may easily become accident-prone if they are careless with equipment, chemicals, or procedures.

Safety Management Programmes

The overall objective of a safety management programme is to eliminate hazards before they cause any serious accidents.

- There are 10 steps in the establishment of an effective safety management programme:
- Review work procedures and inspect work areas for safety hazards.
- Make departmental heads aware of the nature and variety of hazards.
- Establish a safety committee.
- Maintain accurate safety records.
- Conduct periodic in-house safety inspections.
- Train staff members to implement safety consciousness.
- Motivate staff members to be safety conscious.
- Investigate and analyses all accidents and injuries.
- Practice safety management and monitor follow-ups.
- Review the effectiveness of your own safety management programme.

Three Education of Safety

The safety of employees can be ensured by following the three Es of safety: education, safety engineering, and safety rules enforcement.

Safety Education

Safety programmes and policies can only be effective if the staff are trained to think and act safely at work. The best time to start educating employees on safety is during their induction into the establishment, so that they are well versed in safety rules and policies of the establishment before they start their job. Employees should be encouraged to come up with ideas for inculcation safety into the hotel's methods too, and the best ideas should be put into practice and praised or awarded.

The following should be ensured during training:

- Teaching safe methods, with particular emphasis on area of potential danger and how these can be guarded against.
- Demonstrating the use of safety equipment installed in the establishment, and the location and use of first-aid materials.
- Inculcating in people the ability to recognize the signs of hazards around them.
- Teaching staff the legal implications of non-adherence to safety procedures.

Safety Engineering

This involves the building in of safety features into the structure of the establishment-in the equipment, furniture, and fittings and in their proper arrangement within space. Equipment used by the housekeeping employees should be selected to ensure safety in design.

Safety Rules Enforcement

Rules, when not implemented or enforced are not effective. It is not enough to know about safety themes and procedures, but more important to motivate people to put the knowledge gained into practice. This does not come easily to all employees and therefore needs to be enforced by rule and practice.

Occupational Safety and Hazards Standards

Standard. universal laws on occupational safety and health (OSH) do not exist because of differences in local values and cultures. Therefore, differences in local and cultures. Therefore, different countries have developed their own standards on occupational health and safety management systems (OHMS) just as to their needs. India has published and follows the'IS 15001:2000 Indian Standards on Occupational Health and Safety Management Systems-Specifications and Guidance for Use', which is adapted to Indian needs.

The main emphasis of the system is on classifying work activities, identifying hazards, determining risks, deciding if a given level of risk is tolerable, preparing risk-control action plans, and reviewing the adequacy of action plans.

Job Safety Analysis

The executive housekeeper needs to develop a'housekeeping safety manual' for the use of all housekeeping employees. This manual should explain in comprehensive terms the safest methods of performing each task to accomplish a particular job. This has to be done for all the jobs carried out by housekeeping staff. For this purpose, the housekeeper, with the help of the supervisors, needs to carry out a job safety analysis. A job safety analysis is a detailed report that lists every job function performed in the housekeeping department and lists potential hazards, safe methods, tips, and'how-to's for each task involved in the jobs.

POTENTIAL HAZARDS IN HOUSEKEEPING

Due to the nature of the work performed by housekeeping staff, they may be exposed to many dangerous and unsafe conditions, or hazards, if they are not careful. To reduce safety risks, all employees should be aware of potential safety hazards.

There hazards may include:

- Faulty equipment;
- Damaged flooring or chipped tiles;
- Slippery floors and spills not mopped up;
- Slippery guest bathrooms;
- Cracked or broken glass;
- Cleaning agents left uncapped;
- Non-adherence to instructions outlined in the materials safety data sheets (MSDS) for the use of cleaning chemicals.
- Handling corrosive cleaning agents with bare hands;
- Worn-out electrical insulation or fittings;
- Overloaded electrical sockets;
- Trailing equipment flexes;
- Worn carpets and rugs;
- Cleaning equipment left lying around;
- Unsafe use of ladders;
- Loose stair treads;
- Mixing certain chemical cleaners, causing undesirable/dangerous reactions;
- Cleaning agents dept in unmarked or wrongly marked containers;
- Incorrect use of trolleys.
- Incorrect methods of bending and lifting;
- Unsatisfactory hygiene and sanitation standards; and
- Incorrect posture.

SAFETY AWARENESS AND ACCIDENT PREVENTION

Safety awareness should be an ongoing programme at all establishments The management of all establishments should be aware of the laws concerning safe work environments and should be provided to all staff in order to raise awareness about safety. All employees should be aware of the potential hazards in their respective departments. All heads of departments must ensure that employees follow safe job procedures, correct unsafe conditions immediately, and take adequate time to do the job so that accidents are not caused due to haste.

The executive housekeeper should develop a comprehensive list of safety rules to be followed by all housekeeping employees. This can be part of the'housekeeping safety manual'.

BASIC GUIDELINES FOR THE PREVENTION OF ACCIDENTS

The following guidelines can be followed for the prevention of accidents:

- Always follow instructions when using any cleaning equipment.
- Replace caps on cleaning chemicals immediately and securely after dispensing.
- Label cleaning agents clearly.
- Keep floors clean and dry.
- Place warning and safety signs around the area while cleaning.
- Always dry hands before touching plugs, sockets, and electrical fittings.
- Mark faulty equipment as'out of order'.
- Dispose of rubbish carefully.
- Never place cigarette butts of sharp objects in the trash bag on the room attendants' carts.
- Open and shut doors carefully.
- Clean away broken glass carefully.

PROCEDURES TO FOLLOW IN CASE OF AN ACCIDENT

- When a guest or employee has met with an accident at the hotel, the procedure followed should be as follows:
  - With the help of another person, check if the victim requires any assistance.
  - Report the matter immediately to the manager concerned.
  - Either administer first aid (if you are trained to do so) or get help from trained personnel.
  - Transport the victim immediately to a hospital, if required. If the injury is serious, call an ambulance for the same. Follow all necessary first-aid measures until the ambulance arrives.
  - Fill in accident report form and hand it over to the manager concerned.

FIRE PREVENTION AND FIRE FIGHTING

To understand fire prevention and fire-fighting, one must know how fires are classified.

Fires may be classified into four groups, based on their source of fuel:

- *Class A*: These are fires with trash, wood, paper or other ordinary combustible material as their fuel source.
- *Class B*: These are fires with flammable or combustible liquids as their fuel source.
- Class C these are fires involving electrical equipment.
- Class D These are fires with certain ignitable metals as the fuel source

Table. Format of an Accident Report form for Employees

Accident Report Form (for employees)
Name of the injured person..
Section..
Supervisor..
Date...............................Time of report...
Exten of injury..
Was hospitalization required?
Nature of the accident
Time..................................Place..
What happened/ cause..
....
Witness 1..
Witness2 ....
Supervisor's remarks..
Supervisor's signature

Prevention of Fire

Fires may be prevented if fire hazards are identified and eliminated.

Some unsafe practices that may lead to fires are as follows:

- Guests smoking in bed.
- The hotel not providing sand urns or sufficient and appropriate ashtrays in rooms as well as public areas.
- Using high-wattage light bulbs in lamps.
- Leaving linen chute doors open.
- Storing rags and cloths with residues of cleaning polish still on them.
- Not unplugging electrical appliances when not in use.
- Using faulty electrical equipment or sockets.
- Leaving magnifying materials that are easily combustible.
- Using furnishing materials that are easily combustible.
- Each establishment must conduct fire drills on a periodic basis and ensure that all staff attend thee drills so that they know what is to be done during a fire emergency.

Fire Warning Systems

These may be electrically powered manually operated systems, automatic fire detection systems, or a combination of both. The usual components of such systems are discussed here:

Fire Alarms

These can be set off by smoke detectors, heat detectors, sprinkler systems, or pull stations. The most common types of fire alarms are the ones operated by pull stations located in corridors, lobbies, and near elevators. The pull alarms themselves are red in colour, with a glass panel that needs to be broken to set off the alarm.

Sprinklers

These are found in most hotel establishments, especially in corridors and rooms. They are situated on the ceiling and automatically spray water when the temperature rises above a certain level.

Smoke Detectors

These are set off by smoke. The two types of smoke detectors available are photoelectric detectors and ionization detectors. Photoelectric detectors are alarms triggered off when smoke blocks a beam of light emanating from the detector. In the ionization type of smoke detectors, the alarm sounds when the detector senses a shift in electrical conductivity between two plates.

What to do in Case of Fire Emergency

In case a fire breaks out, follow the guidelines given below:

- Immediately switch on the nearest fire alarm.
- If possible, attack the fire with suitable equipment, remembering to direct the extinguishers at the base of the flames. Do not attempt to fight a fire if there is any danger of personal risk.
- Close windows and switch off all electrical appliances, including fans and lights.
- Close the door to the affected area and report to your immediate supervisor for instructions.
- Carry out instructions-for instance, rouse guests in section and direct them to the nearest fire-escape route. Each guestroom should have the route to the nearest fire escape drawn out a displayed in a place where it is most likely to be seen by the guests.
- Report to the departmental fire representative for a roll call. The housekeeper on duty should check the list (in the form of the duty rosters) of the staff who are on duty so that all those on duty can be accounted for.

- Remain at the assembly point until instructed to do otherwise.
- Do not use the lifts.

FIRST AID

The initial assistance or treatment given to a casualty for any injury or sudden illness before the arrival of an ambulance, doctor, or other qualified person is called'first aid'. The notion of'first' aid itself signifies that the casualty is likely to be in need of'secondary aid'.

PRINCIPLES OF FIRST AID

There are certain important principles involved in first aid. These are listed below.

The first aid provider must:

- Act calmly and logically;
- Be in control-both of himself/herself and the problem;
- Be gentle but fire, and speak to the casualty kindly but purposefully;
- Build up trust through talking to the casualty throughout the examination and treatment;
- Explain to the casualty what he/she is going to do;
- Answer honestly and say so in he/she does not know the answer (that is, avoid giving misleading information);
- Never leave the casualty alone and continue to talk to him/her until the ambulance or doctor arrives;
- Continuously reassure the casualty;
- Never separate a child form its parent or guardian;
- Sent the casualty to a hospital or doctor by the quickest means of transport;
- Always inform the police about serious accidents; and
- Inform the relatives of the casualty.

Do not:

- Touch a wound with your fingers or any instrument.
- Put an unclean dressing or cloth over wound.
- Allow bleeding to go unchecked.
- Move a patient unnecessarily.
- Remove clothing unnecessarily.
- Allow a patient with a fracture or suspected fracture to be moved until splints have been applied.
- Neglect shock.
- Risk burning a patient by using an unwrapped hot-water bottle or other hated object.
- Fail to give artificial respiration when needed.
- Fail to remove false teeth, tobacco, or any other eatables form the mouth of an unconscious person.

- Permit air to reach a burned skin surface.
- Wash wounds.
- Try to reduce dislocations, except of the finger and lower jaw.
- Leave a tourniquet on for over 20 minutes without loosening it.
- Forget to send for a physician.

IMPORTANCE OF HOUSEKEEPING DEPARTMENT

- A health club is a part of facilities of most large hotels especially resort hotels this also include swimming pool and spa facilities.
- Accommodation in hotels tend to be the largest part of the hotel, it is the most revenue generating department, the housekeeping department takes care of all rooms is often largest department in hotels.
- Hotel aims to make environment comfortable and offer specialised service to the guest.
- Hotel offer guest the choice of specialty restaurant, coffee shop. The bar also sells liquors which generate the revenue of the hotel. They are available in banqueting, meeting and private party facilities.
- Hotel offer laundry, dry leaning facilities for guest clothes, shoe polishing facilities also.
- Hotels try to make the ambiance as pleasant as possible by nice colour scheme, attractive furnishing and a well kept efficient staff.
- House keeping is the department determine to a large extent whether guests are happy during stay and in turn mankind they return to the hotel.
- Housekeeping is the department that deals essentially with cleanliness and all ancillary service attached to that.
- Revenue can be generated from conferencing, meeting, seminar etc.
- The fine accommodation and service are provided to the guest so they are pleased with the hotel. The guest satisfaction is its primary object and the hygiene factor must always be present in the hotel.
- The rooms in hotels are offered as accommodation to travellers/ guest as individual units of bedroom. Some interconnected rooms are also made which will be helpful to the guest and families. Many hotels offer suits to the guest.
- The standard plays an important role in the reputation of the hotels. One feels comfortable only in the environment which is clean and well ordered, so cleanliness is important for health foremost also for well being.
- These days shopping arcade also found in hotels.

In hotels major part of revenue comes from rooms, rooms which is not sold on any night losses revenue forever and reason for poor occupancy can be anything like hygiene factor, cleanliness, lack of modernizing etc. hence main purpose is to improve whole appeal of the room. A guest spend more time alone in his room than any other part of the hotel, so he can check up the cleanliness he wishes to as some of the guest are more health conscious these days.

He may check up dusting, in-depth cleaning and losses confidence if properly not done *e.g.* If drawers are not cleaned he may generally won't feel like putting his clothes down. Decent room supplies are service like quick laundry and dry cleaning service shows guest that hotel is considering his comfort and wishes to please him. not only this from the cleanliness of lobby, public area, restaurant, cloakrooms, the state and cleanliness of uniform the guest can judge a lot about hotel. it can be positive or negative judgement we can conclude that housekeeping department contributes greatly to all guest impression of the hotel.

DEFINITION OF HOUSEKEEPING IN A HOTEL

Hotel housekeeping departments can be considered hotel ambassadors because of their dedication and responsibility in maintaining the hotel's image.

Housekeeping performs detailed work in guest rooms and hotel areas to provide a clean, comfortable environment for hotel guests to enjoy. Through cleaning and organizing public spaces, housekeeping departments ensure that what the guests see and experience result in a positive impression of the property.

- *Function*: The primary role of the hotel's housekeeping department is cleaning guest rooms. Housekeeping works closely with front desk operations to communicate when rooms are clean and ready for guests to occupy. Although usually associated with cleaning and sanitizing guest rooms, housekeepers are also responsible for other areas, such as public restrooms, convention space and offices. Housekeeping departments often manage laundry operations, which includes washing linens as well as employee uniforms. At some hotels, housekeepers are responsible for minibar inventory and room service.
- *Types*: Housekeeping departments include a variety of job titles. The department will generally have a director of housekeeping, sometimes called the executive housekeeper. This employee is responsible for managing the department and its employees. Housekeeping departments also have supervisors who inspect work and several types of line staff, including room attendants, laundry attendants, turn-down attendants and public space attendants. In addition, other housekeeping employees run the department office by answering phones and dispatching attendants. In some hotels, the office personnel are responsible for managing lost and found items.

- *Size*: The housekeeping department is often the largest employee department in the hotel. The number of employees is relative to the size of the hotel, and may be a combination of full-time hotel employees and temporary leased labour. The number of full-time employees is generally based on the average number of rooms one housekeeper can clean in one shift. The housekeeping department's full size is based on the rooms per housekeeper per day formula, but daily scheduling depends on occupied rooms and/or special projects.
- *Schedule*: Hotel housekeeping departments operate 24 hours each day, but the majority of employees work during the day. Day shift housekeepers are typically room attendants who begin cleaning rooms in the morning. The housekeeping schedule revolves around guest occupancy, so housekeeping departments are at their busiest between guest check-out in the morning and check-in in the afternoon. Turn-down attendants are scheduled for evening shifts, so they can perform nightly turn down in guest rooms. Laundry and public space attendants may work any shift.
- *Tools*: Room attendants generally use a cart to hold their tools and supplies so they can bring the necessary equipment with them to each room. Carts are stocked with chemicals and cleaning supplies to clean surfaces in guest rooms and bathrooms. The cart also holds a vacuum cleaner, broom and trash bag. Although not brought to every room, carpet shampooers and ozone machines can be brought to rooms that require extra cleaning attention.

ATTRIBUTES AND QUALITIES OF HOUSEKEEPING STAFF

HONESTY

This is a very essential attribute for housekeeping staff, especially Room Attendants, who have access to all guest rooms. Guest belongings, sometimes invaluable are often found lying around in the room. The temptations to thieve are great.

It is only the personal quality of discipline and integrity that checks these temptations:

- *Eye for Detail*: It is one of the greatest qualities that house keeping staff must have. It is with this quality that the finer aspects of housekeeping are taken care of and it is what determines a good service from an average one. This quality enables housekeeping staff to take into consideration the minutest details.
- *Co-Cooperativeness*: Housekeeping staff needs to be co-operating with other departments to achieve more efficiency.

- *Briefing and Scheduling of Staff*: Briefing is that process at the beginning of a work shift which is provided by management to facilitate a two-way communication between management and staff. It is the one time during a shift that all housekeeping staff are together to share information and feelings before they disperse to their work areas.
  - Briefing Schedule:
  - Personal hygiene and grooming.
  - Any new policies and procedures introduced by managcl1enL must be made out and explained to the staff.
  - At n briefing the duties or each staff member and the areas of accountability are explained. This would mean that staffs are assigned a floor and allotted their number of rooms. The staffs likewise, are told which floors or public areas they are assigned to for cleaning. They are also told which supervisor would be in charge.
  - Briefing is a time which can be used as a training opportunity. Simple tasks may be demonstrated so that they can be practiced under supervision at their work place.

CLEANING A ROOM

Prior to reporting on a 1loor the room attendant already knows the status of a room in her given lot of rooms. The room attendant can prioritize room to be attended to first on the basis of immediate needs. Before entering the room knock at the door firmly with the index finger knuckle announcing clearly" Housekeeping". When there is no answer, repeat the knock after ten seconds announcing yourself as before. If there is still no answer open the door and knock announcing inside the room "Housekeeping". When there is no reply and one is relatively sure that there is no one in, open the door wide and keep it that way till the entire cleaning cycle in the room is complete.

ROOM CLEANING PROCEDURE

- And centre sashes or windows, close windows.
- Arrange furniture if necessary
- Check the maintenance requirements and report to the department head.
- Clean all the surfaces in single circular motions with a dry cloth. Use a hand dust pan to collect any unwanted matter on the surface without lifting dust in the air Ensure that all surfaces are spotlessly clean. Pay special attention to nooks and corners especially those points that may not obviously be visible to the guests eye.
- Clean lamp shades with a clean dry duster.
- Clean mirror with a dry cloth.

- Clean the bath room.
- Clean the ceiling and air-conditioning vents for cobwebs.
- Collect all loose papers or magazines and stack them neatly on the desk.
- Disinfect telephone mouthpiece with Dettol. Wipe balance of the telephone with a damp cloth, check phone for the dial tone.
- Dust both sides of all room doors, baseboards, window sills, inside and out, bottom
- Dust closet, shelves, hangers and rods. Brush the closet floor.
- Have a last look at the room referring to the checklist for completion of work.
- Switch off the room air-conditioner. Draw all curtains and open the windows for airing the room.
- Switch on the air-conditioning.
- Turn the mattress side-to-side on succeeding days followed by end-to-end turning. Smooth out mattress to air it.
- Use a stiff upholstery brush or vacuum cleaner on upholstered furniture arms, backs and seats.
- Wash the floors.

CLEANING A BATHROOM

- Clean mirror first with dry cloth then with glass cloth and finally with dry cloth.
- Clean the ceiling and air-conditioning vents for cobwebs.
- Cleaning activity starts from the ceiling downwards to the floor.
- Collect all trash in bathroom waste basket and deposit in trash hamper of the maids-cart.
- Finally close the windows shut all lights and close the bathroom doors.
- Floors are cleaned from the wall farthest to the door to the exit.
- Open all exhaust vents.
- Scrub the floor with the prescribed mop and ensure it is dry
- Scrub the toilet bowl and bidet using the special brush or mop and the prescribed sanitizer. The inner rim should be cleaned. Ensure it is dry and spotless inside. Clean the WC from the outside with a sponge till it is sparkling find dry. Clean the lid and toilet seat of the toilet bowl dry and close them.
- Scrub to dry the area next to the wash basin.
- Wipe down tile walls using a sponge or damp cloth. Follow with a dry cloth ensuring that tiles are free of water marks.
- Wipe dry the shower curtain with a sponge.
- Wipe off light bulbs and shades with dry cloth. Check that all bulbs are working.

DIRTY DOZEN

- Air-conditioning ducts and diffuser grills. Under bathroom counters.
- Area above racks
- Behind the WC bowl- the s-trap.
- Beneath the table
- Faucet nozzle filter
- In the toilet roll niche
- Interior surface of drawers
- Rear surface of doors
- Toilet vents
- Top of the door edges and ceiling.
- Top of the picture frames

ROOM MAINTENANCE PROCEDURE

As hose keeping has contact with all rooms and public areas practically in every shift, it is they who detect report and ensure the completion of all maintenance work. This requires close co-ordination with the engineering department.

WEEKLY CLEANING PROCEDURE

In addition to daily cleaning routine, a room attendant normally has some cleaning chores that are of a time-consuming nature. Such items may sometimes be beyond the physical capacity and need the assistance of a skill worker.

Such tasks are:

- Cleaning of window panes
- Polishing of brassware
- Scrubbing of balconies and terraces
- Scrubbing of bathroom tiles
- Vacuuming of under heavy furniture.

FUNCTIONS OF THE HOUSEKEEPING DEPARTMENT

The extension of the housekeeping function outside the hotel bedroom normally includes cleaning of the bedroom floors and may also include the staircases, public cloakrooms and other public areas of the hotel. Many a times outside agencies are involved in cleaning the public areas such as the lobby and the restaurant area as well as the men's and women's restrooms. This activity is normally carried out at odd hours or at the times when public is not around this area. The standard of the cleanliness needs to be maintained and it's the duty of the executive housekeeper to see that pre-determined standards of cleanliness are maintained in all the areas of the hotel it be if the fire-exit stairwells and goods lifts or any other public area. To ensure this, the day of executive housekeepers begin with a round with assistant housekeeper of all such areas which is visited by the public. For this standards need to be set and once the standards are approved, the department shall strive hard to maintain them.

Other function areas of the housekeeping department includes:

- *Lost and Found Department*: Many a times while cleaning a guest room attendant comes across some personal items of the guest left behind. The value of the personal item can vary from almost nil to thousands of rupees as in the case of Rolex watches, etc. All the objects found while cleanin g the rooms are, thus, handed over by attendants to the floor supervisor, who maintains a list of the found objects against the specific room number.
- *First Aid*: First aid as and when needed by the guests is provided by the housekeeping department which m aintains the first aid kit.
- *Floral Arrangement*: Fresh floral arrangements all over the hotel are maintained by the housekeeping department. The private areas like guest bedroom or the public areas like the restaurants and the lobby floral arrangements are made by housekeeping department.
- *Laundry Services*: The laundry services extended to the guest are usually provided by the housekeeping. The laundry to be washed is collected from the guest room and washed, dried and ironed clothes are supplied back to th e guest room. Earlier, all the hotels had in house laundry services but nowadays many hotels have out-sourced this service, both for the hotel linen as well as for the guest laundry.
- *Purchasing*: This is one role of the housekeeping department where a lot of money transactions take place. This is one of the important areas which can affect the profitability of the hotel if various products purchased and supplies are not well managed. The primary areas of purchasing that are under the control of the Executive housekeepers are:
  - Bedroom supplies: *e.g.*, laundry bags, breakfast, cards, shoeshine, etc.,
  - Bathroom supplies: *e.g.*, toiletries,
  - Linen: bed sheets and towelling,
  - Tea and coffee making facilities: *e.g.*, tea/coffee sachets, sugar, milk and biscuits,
  - Working replacements: *e.g.*, shower curtains, crockery, glassware, vacuum cleaners, trolleys, etc.,
  - Cleaning supplies: *e.g.*, air freshener, bleach, cleaning product, and
  - Uniforms: *e.g.*, for the room attendants/porters.

It is necessary for the Executive housekeeper to ensure that the supplies and their consumption is monitored.

The list of items in the purchasing list can be further classified into:

- *Consumable*: Bedroom supplies, cleaning supplies and TCMF
- *Assets*: Uniforms, working replacement and linen.

The Executive housekeeper would have different policies for these two distinct groups of purchasing. The actual consumption of the first group of purchases can be anticipated in line with the hotel occupancy. It is necessary to maintain a stock of these items so that the staff can put up their requisitions on a daily or weekly basis.

The Executive housekeeper is thus required not only to ascertain the demands but also to control the use of consumable products by maintaining properly the invoices and requisition slips. The consumable products are many and if not properly monitored will obviously affect at large the profit statement of the hotel. Today, computers are used for the purpose. "Assets", though not in true sense, like uniforms, etc. are not regularly used up and they do not need to be replaced daily.

But it is the duty of the staff to maintain their uniform and turn up immaculately, daily. Other items like vacuum cleaners, crockery are phased out over a period of time and are recorded as depreciation. Hotel linen is very difficult to manage, as in a large hotel almost 5000 soiled linens are sent for cleaning daily.

To count and counter check the linen from and to the laundry is a herculean task. Moreover, guests also tend to pilfer items like towels from hotels. To keep account of stock and maintain it and when necessary replenish it becomes the priority job for the housekeeping manager.

COORDINATION WITH ENGINEERING DEPARTMENT

The housekeeping control desk have to coordinate with the engineering department for maintenance request that the room attendants registers while servicing the guest room or in the floor.

- The housekeeping department depends on maintenance to keep things in order.
- While carrying out their scheduled work, housekeeping employees may find some deficiencies in the hotel facilities, such as faulty electrical plugs, dripping faucets, leaking pipes or malfunctioning air-conditioning units etc.
- A need for urgent repairs is reported to maintenance over telephone and these requests are usually taken into action immediately.
- There are various heads under which maintenance work is done they are:
- *Electrical work*: Air conditioning and heating; fused bulbs, lights and lamps that are not functioning; defective plugs and plug points; short circuits; and faulty geysers, refrigerators, and minibar fall under this category.

- *Boiler work*: This is necessary to maintain a supply of hot water to guestroom.
- *Mechanical work*: This entails repair or replacement of any faulty equipment, such as vacuum cleaners, ice-cube machines, and so on.
- *Plumbing work*: This deals with faulty faucets (taps), showers, drainage systems, water closets, and so on.

All maintenance requirements needed on floors are entered in the maintenance register kept in the control desk. The control desk attendant notes down the room number, the maintenance work required and the name of the GRA or the supervisor who called attention to the problem.

The desk attendant prepares a maintenance slip in duplicate. She retains the second copy in her book and forwards the first copy to the engineering department.

The engineering department then prepares a 'work order slip' and sends the concerned technician directly to the floor. When the job is completed, a copy of tradesperson's completed 'work order ' is sent to the executive housekeeper for acknowledgement of work completed satisfactorily. If this copy is not sent to the executive housekeeper within an appropriate amount of time, housekeeping issues another 'work order', which signals maintenance to provide a status report on the request repair.

FILES AND RECORDS OF THE DESK

Many important forms, formats, records, and registers are maintained at the control desk.

MAINTENANCE REGISTER

This register is used for recording all the maintenance work required in rooms. Based on the information contained in the register, the control desk attendant fills out the work order form to be sent to the maintenance department.

Table. Maintenance Register

Maintenance register Date:						
Room Number	Time Complaint Lodged	Nature of Complaints	Lodged By	Received by	Time of Completion	Signature

WORK ORDER FORMS

The work order forms are used by the control desk to initiate scheduled maintenance in guestrooms and public areas. A sample work order form is illustrated below:

KEY CONTROL REGISTER

This is one of the most important registers maintained at the housekeeping control desk. It is a part of the key-security system to be followed by the housekeeping department. Each employee who is handed over a key, any key, from key cabinet is supposed to sign for it in a key control sheet in this register. The format of a key control sheet is illustrated below:

LOG BOOK

Another important register kept for reference at the housekeeping control desk is the log book. The log book is used to record all messages that staff from an earlier shift want to convey to the employees on the next shift. All supervisors reporting for work should use the log book for any important messages left for them by the staff of the previous shift. The format of the log book is illustrated below:

ROOM CHECKLISTS FILE

A floor supervisor checks each room prepared by the room attendant, before the room is handed to front office for sale. She uses the room checklist to guide her to examine as per the standards set by the management during her inspection. She ticks the items found okay and makes comments on things which are not upto the standards. The deficiencies have to be rectified by the room attendants immediately. The checklist reflects the performance of the room attendant as well as the supervisor, it is handy to refer to it in the event of a guest complaint. All room checklists are deposited by the floor supervisors at the control desk and filed for a month.

GUEST MESSAGE REGISTER

The housekeeping control desk also acts as a point of contact for in-house guests who require any housekeeping related services. The housekeeping control desk is responsible for taking these guest messages and passing them onto the concerned staff.

The message could be about the provision of certain guest loan items or a request to a second service additional blankets, fresh towels, maintenance requirements etc. A guest messages register is maintained for this purpose at the control desk. The format is illustrated below:

BABY SITTER REGISTER

Babysitting is provided as a service by most hotels' housekeeping departments for guests who have small children. The guests requiring the service contact the housekeeping control desk and the desk attendant enters the request in the babysitting register.

LEAVE APPLICATION FROM

Leave application forms are stocked at the control desk so that they are easily accessible to employees who wish to take leave. The format of a leave application form is illustrated below:

MEMO BOOK

This contains records of all the pending maintenance work for which the housekeeping department initiated work orders. This information is made in copies so as to alert the concerned housekeeping supervisor that work is incomplete. The format of sheet in the meme book is as illustrated below.

LOST AND FOUND SLIP/LOST AND FOUND REGISTER

'Lost and Found' is a term used in hotel terms for those articles left by guests or misplaced by guests in a hotel. Such articles can range from jewellery, costly electronic goods, and travel documents to simple garments. The hotel is obliged to protect such items and return them to the guests. Whenever such items are found by the room attendant or any staff of housekeeping they have to directly report to housekeeping control desk where lost and found is filled. The format of lost and found slip is illustrated below

ACCIDENT BOOK

This records all the accidents of any sort that employees or guest have met with at the hotel. The format is illustrated below, Other Files and Registers that are maintained by the housekeeping control desk are

ROOM OCCUPANCY REPORT FILE

All room occupancy reports are filed for the future reference. The room occupancy reports are important to Executive Housekeeper to determine the level of workload anticipated so as to provide the necessary staff to meet the efficiency each day.

DUTY ROASTER FILE

The duty roaster is filed for information, if required by any one in the department.

7

Office Management in Hotel Industry

A RESERVATIONS MANAGEMENT IMPLEMENTATION FRAMEWORK

So far, this chapter has identified four separate eras of IT development and identified how reservations management either has evolved or may in the future adapt to reflect changes in the IT and business environment. Such 'evolution models' have been critiqued for being based on the linear dimension of time, which means that they imply a strict sequence of processes, changes and activities. In reality, change may be more chaotic than implied.

In proposing alternative models of the reservations process to show how leveraging IT may achieve strategic goals, it should be made clear that each era identifies a state of IT, which may be different across countries, industries or firms at any moment in time. The value of such a model is twofold. In strategic planning terms, it suggests how technological developments could be adopted and harnessed to achieve competitive advantage. In operational terms, it illustrate how systems, operational policies, organisation processes and structure should be configured and aligned for ensuring successful strategic implementation.

So far we have suggested that the technology and what firms use it for go hand in hand. Hence we have described four eras of technology in which the hotel industry has adopted or may adopt four different approaches to reservations management. However, the situation is more complex than that. For as the technology advances, it becomes possible to manage not simply one strategic goal, such as revenue maximisation, but several goals simultaneously. So by the time we reach the 'Content Era' the goals include achieving adding value, cross-selling, customer satisfaction, loyalty, and repeat business. And the technology is not just about reservations but also integrated with marketing, point-of-sale, and other systems.

This is because effective technological implementation is based on two concepts. Venkatraman (1994) suggested that IT use brings more benefits as its exploitation by a business moves from localised exploitation, to internal integration, to business process redesign, to business network redesign and finally to business scope redefinition. The second concept emerges from issues regarding knowledge management and the use of IT to increase the intellectual capital of firms (Nonaka and Takeuchi, 1995). Knowledge management advocates the use of IT in order to develop a digital nervous system (Gates 1999) that would support the structural capital of a company, such as the firm's policies, information database and culture, and which in turn would help to enhance the business' human capital.

Generally, hotel firms have to make two decisions when using IT for reservations management - the level of strategic implementation and the sophistication of its use at each level (Figure 4).

IT can be used as a decision support tool to empower the individual in the reservation or marketing department and to enhance co-operation and integration between departments. It can be used for yield management at a hotel level or it can be used for central rooms management. At each level, the sophistication of use could differ between segmenting by mass-market segments or by individual aiming at the provision of customised services to satisfy individual needs.

Each level requires different management and organisation of each of the YM key areas identified in previous models and should be in alignment and complement the hotel's strategic and market positioning.

The vertical axis represents the digital nervous system, which supports the structural capital of hotels. It represents the systems configuration and integrations that can support reservations management implementation at different levels. The horizontal axis moves from data, to information and knowledge.

This reflects the exploitation and enhancement of the human capital of hotels, the employees and customers. The shift from data to knowledge acknowledges the lifetime value of customers, so that employees are oriented towards enhancing information at each contact point, thereby creating customer knowledge and so more clearly satisfying needs.

An idealised development of both the digital nervous system and what it processes, but that in reality firms may invest one of these more than the other. Some firms focus on the technology, but do not fully exploit it. For instance, Jones (2000) identified that it was up to twelve months before a YM system might be achieving the results that could be expected of it.

Others focus on customer relationships, with limited investment in sophisticated IT. A major reason for these different strategies is that IT investment and systems implementation is typically not smooth but "lumpy". Hotel firms typically review their technology and systems between every two to five years. Implementation is therefore rarely unidirectional and as focused as the model suggests.

RESERVATIONS MANAGEMENT: CHANGING THE RULES OF THE GAME

The term 'reservations management' is rarely used in the literature these days. The term 'yield management' has superceded it. However the two terms are not interchangeable. Reservations management is about process (how to do something) whereas yield management has an output focus (what to achieve). In strategic terms the former is about implementation, whilst the latter is concerned with strategic goals.

This lack of clarity about yield management explains why many definitions of YM are inadequate when trying to understand implementation. Most of them share similar features because they are rooted in Kimes' original definition (1989), which was:

- "A method that can help a firm sell the right inventory unit to the right customer at the right time and for the right price. It guides the decision of how to allocate undifferentiated units of limited capacity to available demand in a way that maximises profit or revenue."

But this definition does not differentiate between pre-YM reservations management and contemporary YM practice. Hotels have *always* tried to sell the right room at the right price to the right customer. This is a fundamental goal which remains unchanged, whatever means are used to achieve it.

The point is that the *process* of yield management is a new approach to reservations management because it enables hotels to achieve this obvious goal in ways that they never could before. Jones (1999) proposes a definition based on process not output, which explains these key elements of a YM system as follows:

"Yield management is a system for hotel owners to maximise profitability through their senior management in hotels identifying the profitability of different markets segments, setting pricing, discount and displacement rules for application to the advance reservations process, and monitoring the effectiveness of these rules and their implementation."

What this definition does not explain is that setting rules and applying them is only made possible by new technology. IT and YM systems are able to handle and manipulate data in new ways so that forecasting and decision-making are greatly improved. In the rest of this article we shall explore in some detail how this has been made possible, and we attempt to show how the 'rules of the game' have changed and will change in the future.

TECHNOLOGY SHIFTS AND RESERVATIONS MANAGEMENT PRACTICE

IT has undergone dynamic and rapid development. Such change is driven from two sides. On the supply side, new technologies are emerging fast, whilst the business environment increasingly demands more IT support. There are two well-established frameworks concerned with the development of IT in organisations. Both of these provide insights into the ways in which IT is evolving, offering management the possibility to exploit this complex phenomenon, without going into the technical details.

Nolan (1979) proposed a model describing the growth processes and the development of the use of IT in organisations composed of two eras, namely the Data Processing (DP) era and the IT era. Mutsaers et al (1998) extended this model by adding the Network era (Figure 1). Each era has its own characteristics in both business and IT terms. The transition from the DP to the IT era is typified by a *technological* discontinuity while between the IT and the Network era an *organisational* discontinuity can appear.

The descriptions and analyses of the different eras, proposed by these two frameworks, suggest that they have considerable overlap. What follows is a necessarily simplified overview of the evolution of IT use, within the context of hotel reservations and its implementation. The analysis follows an evolutionary model composed of four eras combining the two previously mentioned frameworks.

THE DATA PROCESSING ERA

The dominating principle through most of this period was that of Grosch's law, which stated that computer power increases as the square of the cost. Thus in the 1970s it became economic to process and analyse large complex data sets quickly and accurately by using IT. The main characteristic of this era is that firms really began to pay attention to automation. The business environment in the DP era is relatively stable. Businesses are typically functionally oriented hierarchies who use IT mainly to improve efficiency, often by replacing humans with machines.

In the hotel business, the power of IT was used to automate the reservation process and to set up central reservations systems. In essence data were recorded electronically, but little or no shift occurred in advanced reservations practice. Whilst computers were able to store large amounts of data, the software based on yield algorithms did not exist to enable the effective management of yield.

In this era, strategic competitive advantage derived from ensuring the technology delivered a lower cost of reservations, facilitated referrals across the chain, and enabled monitoring of marketing campaigns.

PC-CENTRIC ERA (THE DEVELOPMENT OF UNIT-LEVEL YM)

The underlying IT economics of this era were summarised in Moore's law, which stated that semiconductor price/performance, would double every two years for the foreseeable future. This prediction remained fairly accurate through the 1990s, helped by the constantly improving designs and processing capacities of proprietory micro-processor-based systems. Many peripherals and PC software tools took on commodity-like characteristics. This resulted in a rapid and wide technological diffusion across all hotel operations, in the form of personal computers, new types of personal software (such as spreadsheets and word processors) and the expansion of chip technology in point of sale terminals, data communication networks, and work stations.

For hotel reservations, IT was now able to engage in demand analysis and forecasting, and to carry out complex calculations on large data sets. 'Embedded knowledge' enabled reservationists to accept or decline reservations very simply.

Management was responsible for the strategic decisions in determining pricing policies and the hotel's market positioning and market segmentation. Management was also responsible for monitoring YM implementation, controlling its effects on profit maximisation and aligning reward schemes with profit increases.

Jones (1999) suggests that the reservations system has two principal sub-systems - a decision support system and a decision-making system that is further sub-divided into strategic and operational decision-making (Figure 3). In the decision support system, demand analysis and reservations compose the core of the system on which the decision-making activity is based. This is supported by both a technological and a human resource input.

The external environment affects both YM systems and their subsystems. Although the strategic planning system makes market segmentation and pricing decision on an annual basis, as in the previous era, such decision are implemented far more flexibly at the unit level. The technology, or rather the YM system, delegates authority for goal achievement to the unit. In this era, competitive advantage derived from managing this complex system effectively, as accurate forecasting and the application displacement rules could lead to revenue enhancement of up to 8% (Jones 2000).

But not only was the reservations process improved, this technology also led to additional strategic advantages. Technical developments in distributed computing architecture, together with organisational reactions against local, costly, frequently inefficient microcomputer-based initiatives helped firms to organise according to markets, product lines or geographic areas. There was also a shift towards open systems, with the promise of common standards and high compatibility.

Despite the difficulties and the underlying costs of such integration, this facilitated system interfaces and data sharing. In the hotel industry, systems such as distribution systems, electronic point of sale (EPOS), property management systems (PMS), and reservations were able to communicate and exchange data. This led to tighter control of operations through improved co-ordination and reduced the cost of such co-ordination.

THE NETWORK ERA (CENTRAL ROOMS MANAGEMENT)

Though the Internet has existed for nearly 20 years, it was the arrival of the Mosaic graphical interface in 1993 that allowed access to mass markets based on the Internet and the world wide web (WWW). This era is defined by the integration of worldwide communications infrastructure and general purpose computing. Communications bandwidth, restricting as it does the graphical capabilities of the WWW, begins to replace micro-processing power as the key commodity. Attention shifts from local area networks (LANs) to wide area networks (WANs), particularly intranets. Economically, Metcalfe's law replaces the pre-eminence of Moore's law. This law states that the cost of a network rises linearly as additional nodes are added, but that the value can increase exponentially. Software economies follow a similar pattern. Once software is designed, the marginal cost of producing additional copies is very low, offering huge potential economies of scale to the supplier. Combining network and software economies produces significant opportunities for value creation. By the late 1990s, the fundamental network-centric applications were email for messaging and the WWW for e-commerce activities, which promise even lower transaction costs.

Indeed, the arrival of the WWW added significantly to the distribution channels available to hotels. Hotels could now promote directly to customers and sell their rooms inventory more cheaply than using the expensive call centres, global distribution systems (GDS) and central reservation systems (CRS) of their hotel affiliation. The arrival of digital TV and the wireless application protocol (WAP) has further increased the number of available distribution channels.

The increase in both the number of and the interconnectivities between distribution channels has made the maintenance of a single-image room and rate inventory a very cumbersome and difficult task for hotels. As a result, hotels argue that they have lost control of their inventory, while customers are frustrated and confused by the fact that different channels quote them different rates and room availability (IHRA, 1999). It is also increasingly difficult to sell the last available room and very often under or indeed over bookings can occur. The negative effects on hotel revenue and customer satisfaction are clearly obvious.

In an effort to address the problem of room and rate integrity, hotel chains are increasingly transferring the practice of YM from property level to chain level (Connolly, 1999). In addition, the tendency towards centralised management and control of multi-unit hospitality units also leads to the concept of central rooms management (RM). Central RM occurs where the revenue analyses of all market segments for a given group of hotels are performed at one central site, not at the individual property level (Bennett, 2000). Central RM has several advantages. It considerably decreases the maintenance and software costs as well as creating a central focus for a successful single-image system. At the same time, in a highly competitive environment where brand differentiation becomes critical, hotel companies must ensure that their pricing philosophies are consistent across the range of brands. Central RM puts this in the hands of one department and possibly one place, not in the hands of dozens or even hundreds of individual revenue managers across vast geographical distances. Hotel chains can now manage yield across a geographical area and not just at the property level. As a dizzying number of new and alternative distribution channels arrives on the scene, central RM may position an organisation to control those channels better and to respond to them in the most profitable way for the brand as a whole. Very soon, central RM systems will be more accessible to mid-level and smaller chains, as the application service providers (ASP) model, or the single-image inventory, becomes more prevalent across the hospitality industry.

Such solutions are already available from IT vendors. Several reservation provision companies are moving towards co-operation with YM vendors (Hotel-online.com, 2000). For example, Integrated Decisions and Systems Inc. (IdeaS) are promoting their 'e-yield™' software (http://www.idealyield.com/solutions_n.html), which provides a network YM solution across a LAN, WAN or the Internet, indeed anywhere a group of properties or a worldwide chain require it. In the near future 'e-yield™' will deliver benefits such as on-line pricing and availability and will integrate directly into a hotel or chain's internet marketing and distribution capabilities.

Dynamic or real time pricing models, whereby rates change continuously like a stock market depending on the supply and demand conditions, in effect customers bidding for rooms, has also been proposed in the literature, (Davis and Meyer, 1998). This reflects the nature of the emergent patterns of customer purchasing behaviour on the internet.

In this network era, reservations management is even more a strategic implementation tool, with control at head office. The corporately determined strategy towards distribution through multiple channels with centralised management and control of hotel units is reinforced. Demand is not analysed by market segments rather, it is analysed at a finer level of analysis concerned with the purchasing behaviour of different segments through different distribution channels against the demand conditions of properties in different locations operating under different brands.

Demand is forecast for each distribution channel, hotel brand and location and YM practice directs sales towards the most profitable and least expensive channels as well as facilitating the cross and up selling of customers between different locations and/or brands. Competitive advantage derives from the firm's ability to leverage its brands across these multiple distribution channels.

CONTENT ERA

Mochella (1997) argued that the key shifts in the content era will be from electronic commerce to virtual businesses; from wired consumer to individualised services; from communications bandwidth to software, information and services; from on-line channels to customer pull; and from a converged computer/communication/ consumer electronics industry value chain to one of embedded systems. A content-centric era of virtual businesses and individualised services would depend on the previous era delivering an inexpensive ubiquitous and easy to use bandwidth infrastructure. The economic driver would be 'information economics', combining the infinite scale economies of software with the nearly infinite variety of content. The law of transformation would supercede Metcalfe's law. Here, a fundamental consideration is the extent to which an industry or business is 'bit-based' (information led) as opposed to 'atom-based' (reliant on physical product). In the content-centric era, the extent of an industry's subsequent transformation would be equal to the square of the percentage of that industry's value-added accounted for by 'bit' as opposed to 'atom' processing activity.

What does this mean for the hotel industry and reservations management? Competitive advantage would no more be sustainable based on the simplistic "location, location, location" argument of the first era, or "yield management" of the second, nor "brand ubiquity" of the third. It will be dependent on the creation of "knowledge, knowledge, knowledge" in order to provide added value to customers (IHRA 1999). The physical inventory of rooms would be replaced by distribution systems and networks, which would drive value creation.

As customers become more knowledgeable, demanding and require a personalised product, hotels would rely more on their guest information assets and knowledge technologies, such as artificial intelligence and intelligent agents, to enhance further their 'mass customisation' and one-to-one marketing strategies.

Several hotel chains already implement customer relations management and one-to-one strategies by changing the way guest information is collected, analysed and used at every interaction with a customer. Personalised web sites, sales offers and information are distributed through the internet or other internet enabled devices, such as palmtops and telephones.

The basis for one-to-one marketing is share of customer, not just share of market. Instead of trying to get as many customers as possible, the focus shifts to gaining a greater share of business from existing customers. It highlights the importance of understanding the lifetime value of a customer and building long-term relationships.

These changes have led to a general business concept called 1:1 YM™. What distinguishes 1:1 YM™ from traditional YM is the focus on the specifics of an individual customer's situation and the value of the overall customer relationship versus focusing on general customer segments and the value of a specific customer's discrete transactions (Karadjov and Hornick, 2000).

In the former models of yield management and central rooms management, the optimisation process considers the value of each transaction as a customer requests it. The value is usually taken to be the revenue or, in more sophisticated models, the contribution generated by that transaction.

In other words, a typical YM model is oblivious to the characteristics of the customer, and views the sale only as a price or contribution typical of the segment they represent. 1:1 YM™ brings one-to-one marketing and YM techniques together at the point of contact with each customer. It informs the price or availability decision with an insight into who the customer is, what the specifics of their current situation are and what is the overall value of their relationship. Hence, a hotel manager may decide to set aside rooms for high-value, long-term customers who stay often or generate significant ancillary revenue at a particular property or with other properties from the same chain. These decisions may involve complex tradeoffs between present revenue/ contribution and potential future business.

Orkin (2000) argued that 1:1 YM™ can not be done from a property-based system because it needs to manage yield on the total customer history and all ancillary spending. It needs the central YM concepts brought forward by developments in the Network Era. He also argues that 1:1 YM™ evaluation cannot rely on traditional YM metrics based on profit enhancement. New control and evaluation systems should address both the issue of the customer's lifetime value and customer loyalty, as well as their importance to hotel revenue.

This means that incentives would also have to align with such changes. Indeed, the hotel's 1:1 YM™ culture should be based on understanding customer needs and reinforcing customer relationships. Thus, all customer-facing processes and contact-points would have to be redefined and employees rewarded for assisting in customer data collection and customer needs fulfillment. 1:1 YM™ may also be the answer to the commoditisation, i.e. sales based on price only, of the hotel products featured by several infomediaries on the Web, e.g. priceline.com, lastminite.com, Ebay.com (IHRA 2000).

PERSONALITY CHARACTERISTICS OF HOTEL MANAGERS

The third predominant research area, which draws from psychological theory and has received substantial treatment in the hospitality literature, consists of studies that focus on the personality characteristics of hotel managers. Although it is not clear why so much research has been undertaken in this area, it is perhaps because the popular perception of hotel managers is that they are a particular personality type. Arnaldo provided a profile of hotel managers and identified the personal characteristics of general managers, and the general managers' allocation of time and importance to Mintzberg's ten managerial roles. The study examined general managers' education, job satisfaction, effective measures of success and the future for the managerial training. Kim provided a profile of hotel general managers in Korea, again using Mintzbergs' managerial roles. Arnaldo examined the general managers' allocation of time to the ten Mintzberg managerial roles. In addition the study identified the demographic characteristics of the general managers, and aspects of the general managers' job satisfaction. The paper makes recommendations for the future training of managers, such as increasing training in personnel management, accounting and finance.

The managerial roles of leader, monitor, disseminator and entrepreneur are judged to be important, and it is crucial they are included in training programmes. Brownell examined the gender differences of hotel managers in relation to personality and career development. The focus of the study was on how the personality traits of hotel managers influence career development in the hotel industry. The study identified that women are more in agreement than men about the characteristics they believe contributed to their career development. Most often mentioned by women are interpersonal skills, determination and hard work. Men responded with integrity, hard work, and interpersonal skills. The diverse backgrounds of hotel managers were also identified by Bentivegna and Sluder, who give a detailed demographic breakdown of a sample of 237 hotel general managers in the USA.

Worsfold examined the personality profiles of a group of UK hotel managers. The personality profile of hotel managers are described, and differences are identified between the personality profile of hotel managers and managers in other industries.

Worsfold compared the managers' self ratings against their personality profile, as measured through Cattell's 16PF test. Compared with the general management norms provided by Cattell, the personality characteristics of the managers are shown to be more assertive, more venturesome, more competitive and more tough minded, more independent and to have lower levels of anxiety than Cattell's management sample. Again, the trend towards showing that hotel managers are unique and different from other managers is evident here.

Swanljung's paper is one of the few studies published that attempts to identify common personality characteristics to the success of hospitality managers. Fourteen successful North American hospitality executives were interviewed and asked to identify their personal characteristics. Swanljung identified that the hotel executives have to be determined, hard working, fair, and able to motivate others. Stone also replicated this study.

Here personality characteristics are related to effective hospitality management using a survey of 140 UK general managers. Stone postulates that hotel managers demonstrate particular forms of behaviour, and identifies a number of factors that identify them from non-hotel managers. For example, compared with other managers, hotel general managers are more assertive, stubborn, cheerful, competitive, active, independent, cynical, calm, socially bold and spontaneous, harder to fool, and more concerned with self. Seymour in a study of 30 managers of small hotels in France found that the first thing in common with all the managers was their belief that they were somehow different or individualistic.

In a similar study by Nebel the background and the personality characteristics of a small sample of ten general managers are examined. The study focuses on five separate categories of personal characteristics: needs and drives, attitudes and values, interpersonal orientation, temperament and cognitive orientation. The study concludes that although the personality characteristics of successful hotel managers are varied, there is evident consensus of what characteristics are required in order to successfully carry out the hotel managers' job. What these research objects show are strong commonalities in the sense that somehow personal characteristics are the dominant measure of success in a managers' career, rather than other external variables put forward by authors such as Ladkin and Riley.

An additional use of personality characteristics is evident in the research by Dickinson and Ineson. This research identified personality characteristics by the use of biodata as a means of assessment in the selection of operative staff in hotels.

The authors conclude that the use of biodata to determine personality characteristics holds potential for the selection of reliable operative staff whose job involves a degree of customer interaction. Finally, the personality characteristics and profiles of hospitality management graduates in the USA were detailed by Brymer and Pavesic. This research attempts to show through the use of personality traits why some graduates remain in the industry, and why others take the decision to leave. The study concluded that quality of life issues and academic preparation are important to industry retention of hospitality graduates.

In summary the literature concerning the personality characteristics of hotel managers has common aims insofar as the research focuses on whether there are particular personality traits that set hotel managers apart from other professions and whether specific personality traits result in career success. The fact that many of the studies are not conclusive means that there is scope for more research here.

However, the research calls upon psychological theory to help better understand the people who become and are successful hotel managers. There is a practical application here for industry, as research identifying the right type of individuals for the job may have benefited in terms of successful hotel management.

THE CAREERS OF HOTEL MANAGERS

The second category within the hospitality literature for exploration focuses on the careers of hotel managers and draws upon the theoretical foundations of career theory. At the macrolevel there are a number of studies that have specifically examined hotel general managers' careers. The first, by Guerrier, examined hotel managers careers and their impact on the cultures of hotel organisations in Britain. A number of characteristics of the hotel industry were identified, for example the insularity of hotel managers who tend to be trained separately from managers in other industries, and the emphasis of mobility within hotel managers' careers, which helps to define the character of the industry.

The paper explored ways of understanding careers; the careers of hotel managers; the hotel industry; and finally, the implications of the traditional approach to career development. Although largely descriptive, the paper provides a useful explanation of hotel managers' careers. As with the nature of managerial work, again the idea that hospitality management is different from general management practices is raised.

Three main stages in a hotel manager's career are identified by Guerrier. First is the training stage, which usually begins with specific hotel and catering education. Second is the assistant manager stage, where the employee is usually given some kind of functional responsibility, the two main functions identified as front of house and food and beverage. The final stage is the general manager stage, characterised by increased mobility as managers move through an increasing size of hotel, which increases pay and prestige. Guerrier argued that the careers of hotel managers are best understood from an occupational perspective rather than an organisational perspective, as managers become moulded to an occupation rather than an organisation.

Finally, the differences between the traditional hotel manager and the modern hotel manager were identified. This study cited the characteristics of the traditional career path of managers as a barrier to the development of female general mangers. Similarly Baum through a wider study of the hotel industry in Ireland gives an account of the career paths of Irish hotel managers, and examined the needs for training and human resource development in the Irish hotel industry.

Building on the work of Guerrier and continuing the theme of career patterns and mobility, a study of the career paths of hotel managers in the UK was provided by Ladkin and Riley. Here the use of age distributions is tested and the notion of career logics is introduced for the first time as a method of interpretation of a career sequence. Issues of job mobility, length of time to general manager, and the salience of specific jobs to career development are explored. Food and beverage was found to be the most salient function in career development.

In updated research the focus moves to concentrate on the number of job moves in terms of whether they are internal or external to the company, and whether the individual or the company initiated the moves. In this research a whole range of variables are tested to see how they influence the shape of a career. An American perspective of the same topic was provided by Nebel *et al.*, who explored hotel general manager career paths in the USA.

A number of key findings are addressed, the most notable being that most general managers have a narrow career path before their first general manager job. A reported 87.5% spent their time in only one department.

Nearly 80% had held assistant manager positions and two major hotel operational departments, 'food and beverage' and 'rooms' accounted for three-quarters of general managers' career paths. It was reported that this trend is increasing, as the figures were higher for younger managers. These studies show that there is still an importance of traditional craft training in hotel management.

The manpower and labour market components of careers are also addressed. A study of career paths of hotel general managers has been explored from a manpower planning perspective by Riley and Turam. In this study of 115 UK hotel managers, the authors identified the length of time taken to reach general manager status and the stages of development involved in progressing to reach this target job. The study produces a number of interesting findings with regard to functions, mobility, and locus of career control. These findings are supported by Guerrier and Lockwood, who distinguish between company core and unit core staff in the hotel industry in terms of positioning in the labour market.

Related to the manpower planning approach is the labour market perspective, which has been explored by Ladkin and Riley. Here there is an examination of the career paths of hotel managers in relation to the bureaucratic model, which dominates career theory. The argument put forward is that the opportunity structure of the industry encourages a form of career management that involves, in addition to jobs and organisations, the labour market as an entity in its own right. The study explored mobility, skills, knowledge and personal planning strategies in relation to career advancement. With respect to career advancement and continuing with the labour market theme, two types of contrasting career paths for hotel managers have been outlined by Beck. This study compared the corporate-paved route and the self-planned route to industry management.

The paper postulates that for graduates who do not join planned traditional corporate training schemes, an alternative route is available to them whereby the graduate works his/her way up through the ranks of the industry. The object concluded by presenting opinions from an industry with regard to the best way for an ambitious graduate to develop his/her own route to management.

Swanljung identified five broadly defined career routes of hotel executives: the operations route, the accounting and finance route, the marketing route, the food and beverage route, and the non-hospitality route. Taking a psychological stance, the perceptions of hotel managers' careers from the perspective of supervisory employees are examine by Williams and Hunter.

This paper examined the transition from being a supervisor to a hotel manager, and identifies the hesitancy and difficulty in making such a career shift. What this collection of papers represents is research that examines career paths and the variables that influence the shape and form of a hotel managers' career. The most recent research draws heavily from career theory, and has a practical application in so far as it sends messages to the industry concerning labour market behaviour, established career routes and skill development.

The subject of labour markets in the tourism and hospitality industry is explored in career research. Two major lines of enquiry are identified. The first labour market issue that has been examined is labour market trends in the tourism and hospitality industry. For example, Boella described a background of the hospitality industry's workforce, and examines changes in the hospitality industry and the industry workforce.

Hornsey and Dann examined the nature of organisations in the hotel and catering industry, and classify the nature of jobs in the hotel and catering industry. Issues of conditions of employment, hours of work and pay are also discussed. Baum reviewed a range of books that focuses on human resource issues, and examines tourism labour markets. Wood researched the British and international experience of hospitality industry labour trends, and Lucas examined recent trends in hospitality employment, and looks at the possible developments in the 1990s. The American hotel and catering labour market trends are reviewed by Tanke. One of the emergent key findings in these reviews of labour market trends is a shortage of labour. This has direct implications for the industry.

Addressing this labour shortage, an examination into core and peripheral employees in the hotel industry was undertaken by Guerrier and Lockwood. This flexible working approach attempted to alleviate problems of recruitment and retention of staff in hotel operations, in addition to responding to pressures for delivering a high quality service, therefore improving the performance of employees. A model of core and peripheral staff within hotels was proposed. Human resource needs and priorities for the hospitality and tourism industry were outlined by Teare and Brotherton and the job opportunities offered to hotel and catering graduates were explored by Jameson and Hargreaves. Problems of seasonality for workers in the tourism labour market have been examined by Ball.

The second labour market issue looks at the subject of labour turnover. Bonn and Forbringer examined the issue of reducing turnover in the hospitality industry, and present an overview of recruitment, selection and retention. Riley presented the relationship between recruitment, labour turnover and occupational rigidity in the hospitality industry. Hornsey and Dann examined labour turnover in the hotel and catering industry, and Johnson focused on labour turnover specifically in hotels.

The labour retention strategies of UK hotel managers are illustrated by Riley. Finally, in a review of labour turnover, Riley identified six themes evident from the literature. These are; the extent of labour turnover, attributed causes, solving the labour problems, labour turnover and labour economics, images and perceptions, and informed opinions. Riley argued for a change in the paradigm from these approaches to one that addresses the relationship between commitment, job satisfaction, and the intention to leave. The idea that the high rates of turnover is a damaging feature to the industry is questioned by Riley, who argues that labour mobility plays an important role is the development of skills.

In addition to the dominant research themes of career theory and labour markets, a number of specific career issues drawing from career theory have received attention in the hospitality literature. The importance of career planning in the hospitality industry was addressed by Antil. The author stated that career planning is a continual process, and stresses the importance of not leaving career opportunities to occur by chance. The main thrust of the argument is the importance of individuals and organisations making the right match. The means of achieving this match is said to be through career planning.

Although not specific to hotel managers, Ross examined management values, work preference ideals and personality needs as predictors of hospitality industry career anchors. Using Schein's conceptualisation of career anchors, these are considered as explanatory variables among Australian hospitality industry employees.

Ross argued that despite the rapidly growing importance of the hospitality industry in terms of economic benefits, little is known about personality and socio-demographic variables of hospitality employees. Murphy and Ross have found that certain groups of people appear to develop much more positive attitudes to the tourism industry in any tourism-related community.

In this study, 274 employees within the hospitality industry in Cairns, Australia were sampled to test their management values, work preference ideals and personality needs in relation to career anchors. Using a range of measurements to define career anchors, the career anchors of job satisfaction, job security and continuing employment were identified as the most predominant anchors.

The major career anchors suggest that many employees' fundamental motivation is associated with feelings of work satisfaction within an employment environment offering both stability and continuity. The implications for hospitality industry management and human resource managers are discussed. Authors drawing from Schein represent key ways the work of a theorist has been applied to the hotel management literature. Similar to Mintzberg, Schein has been adapted by researchers in an attempt to explain personal variables in career paths.

Career development issues have also been covered in the literature concerning hotel managers. Tanke offered a definition of career development, and examined the role and the intricacies of career development programmes for hospitality management. The career development of hotel managers has been researched by a number of authors. For example, Ruddy in a study of 31 hotel managers in Hong Kong addressed the issue of what general managers believe are the key influences in helping them to become successful.

The conclusions drawn from this study are that the managers place qualities related to their behaviour and personal characteristics highest on their list of career influences. The less important factors were external to the managers and consisted of family support, mentor relationships and education. A further study as part of on-going research to focus on the patterns of hotel management development in South East Asia, focused on the factors that have influenced the success of a sample of hotel general managers.

The respondents place qualities related to their personal characteristics and behaviour highest on their list of career influences. The respondents gave an importance to human relations and communication skills when asked what others should do to become successful hotel managers. The author argued that the future direction of research in this area should be to determine what it takes to become a 'high flyer' in the hotel industry.

Dodrill and Riley examined hotel worker's orientation to work, to see if a unique orientation to work can be identified. Five attitudes were measured; ambitions, the desire for scope in the job, mobility, security and autonomy. No distinctive attitudes towards work by hotel workers were found.

Damonte and Vaden undertook a study to examine the nature of career decisions in the hospitality industry by looking at the experiences affecting career decisions, characteristics important in a career, and sources of career satisfaction. The career plateau receives specific attention in research by Reed and Evans. This study urges the hospitality industry to recognise when career plateau's exist, and examines ways in which the industry can respond to the problem of people who can no longer be promoted. The 'career plateau' is a concept taken directly from career theory.

A comprehensive review of the organisational commitment literature as applied to the hospitality industry has been provided by Paxon. The research states that few studies of organisational commitment relate to the hospitality industry in contrast to the number of studies that exist in organisational behaviour and organisational psychology literature.

CAREER PROBLEMS

A further specific topic of investigation with regard to the careers of hotel managers is the problems associated with the job, and the effects that these problems have on hotel managers. In a study by Larmour the problems faced by managers in the hotel and catering industry were examined. The study revealed that managers in the hotel and catering industry rate their problems as different from managers in other comparable businesses. Within this, the problems are different for managers of different sectors within the hotel and catering industry. Again we see the recurrent theme of hotel management being perceived as different from general management.

Specific attention is given to job stress. Managerial job stress in the hospitality industry is examined in detail by Brymer *et al.*. This study stated that despite the large range of research into managerial job stress, little research has been undertaken with a specific focus on the hospitality industry. This study aimed to rectify this situation, and examine three types of stress on middle and upper level managers in the hotel industry in the USA. The results of the study indicate that perceived job stressors are positively related to overall strain, and the use of negative coping mechanisms have a positive relationship with overall strain.

The study concluded that perceived job stressors were related to self-reported psychological strain and behavioural strain. The authors call for stress management programmes and employee assistance programmes to attempt to reduce stress at work, in view of the financial cost to organisations in employee sickness and layoffs. Sarabakhsh *et al.* examined the personal cost of hospitality management by investigating how hospitality manager's working conditions affect their overall quality of life. The study consists of 42 managers in food service operations, 41 hotel general managers and 68 blue cross/blue shield employees. The research reveals that the three measures of life stress; career interference with social life and career interference with a happy marriage all scored significantly higher for hospitality groups. The self-esteem and anxiety ratings are not significantly different amongst the three groups. The disturbing conclusion from this study reports that the situation was much worse for hotel managers, who repeatedly scored the highest in all three stress factors.

The author argues for more emphasis on teaching staff how to delegate and to give managers assistance by not short-staffing organisations in an attempt to alleviate these problems. Brymer and Pavesic examined job satisfaction in an attempt to determine why so many young managers are leaving the hospitality industry. The study states that the concept of the traditional career path has become inconsistent with the living styles of junior executives. The young managers state that although they like the people aspect of the job and the challenge that the career presents, they are not happy with the long hours. The respondents can point to no single motivator for leaving the industry, and are more likely to be unhappy with the job rather than the career. The authors concluded that more help is needed in the critical time three to four years after graduation if the substantial loss of young talent is to be curbed. In a similar study, Reichel and Pizam compared the job satisfaction, lifestyle and demographics of hospitality workers in the USA against employees in four other industry groupings. In all three areas the hospitality workers were different from the other groups. The findings revealed that the employees were less satisfied with their jobs than the other workers.

Although not concerned directly with hotel managers, a number of related specific functions are examined in relation to job stress and manager's burnout. For example, Krone *et al.* examined 'manager burnout' for a sample of food service managers. Using the Maslash burnout inventory of the 532 respondents, the results showed that 20±25% of the managers fell into the high burnout range, over 25% showed high levels of emotional exhaustion, and 33% demonstrated levels of emotional exhaustion in the middle range. The factors cited for this burnout were role overload, role ambiguity, and role conflict.

Whitney examined the attention styles and stress factors of hotel sales and marketing managers. A range of attention styles are presented and the study shows that effective performance depends upon a good match between situational demands and attention styles, which reduces stress and overload.

What these studies have in common is an attempt to understand the problems of hotel management as an occupation, and how these problems can be alleviated. This has obvious implications for industry.

MENTORING

As an attempt to alleviate job stress and career problems, a subject that has received increasing attention in the management literature relating to organisations is mentoring. Research that specifically focused on the role of mentoring in the careers of hotel managers is evident from Roche and Schmidt. In a survey of 200 general managers, Rutherford and Wiegenstein examined findings with regard to mentoring in the hotel industry. A number of variables were examined, for example, the role of a mentor in enhancing career success and satisfaction, and aiding career moves. The authors conclude that although the presence or absence of a mentor cannot be proved to distinguish managers on the basis of job satisfaction, career success or managerial effectiveness, mentoring has a positive influence on the careers of hotel general managers, specifically with regard to career moves and confidence building. Williams and Hunter in their study of supervisory hotel employee perceptions of management careers identified an informal nature of career mentoring for supervisors.

Eighty per cent of the supervisors claimed that they could rely on the advice of a person higher in the organisation with regard to advice on career opportunities, although more than three-fifths of them claimed they were not in receipt of a formal mentoring system devised by the organisation. Mentoring represents one way in which the industry could reduce labour turnover and alleviate some of the difficulties faced by those in the hospitality professions.

To summarise, the literature relating to the careers of hotel managers focuses on a range of issues drawing from career theory, for example, career planning, career development, job satisfaction and career anchors. The issue of labour turnover and job mobility is a recurrent theme that shows the importance of mobility for the industry and in shaping career paths.

It is perhaps in this regard that the hotel manager's literature makes the most of established theory as a means to understand the ways in which managers develop their careers, and the structures and characteristics of the industry that facilitate career development. However, like the hospitality research that comes from management theory and Mintzberg, there is a dominance of only one key theorist, in this case Schein. There is clearly room for expansion. In an attempt to address this, one of the most comprehensive attempts to integrate career theory, hospitality and tourism is provided by Riley and Ladkin, who developed a basic analytical framework for integrating career theory and tourism.

Following an overview of the main areas of research on careers, a framework is devised for analysing careers in tourism using a structural±technical perspective. There are three structural±technical determinants of a career; organisation size and structure, technological specificity, and dispersion of knowledge. In addition, career behaviour affects the career path, which sits within the structural variables. Sequence analysis is suggested as a way to research careers paths, and the variable of the labour market is introduced.

RESEARCHING DEVELOPMENT AREAS IN THE HOSPITALITY INDUSTRY

Areas that offer the most potential for growth should be explored. Because such areas change frequently, it is not possible to list them in a textbook. However, factors that support continued growth and strong business activity are regularly reported in such publications as Trends in the Lodging Industry, by Pannell, Kerr, and Forster. Hospitality industry futures projections publications usually cover such issues as new hotel developments; hotels under consideration; activities of convention and visitors' bureaus; strength of local economies; development of business, recreation, and arts activities; the need for office space; and area hotel occupancy percentages and average room rates.

This information is listed for selected cities both within the United States and at international sites. The Internet is increasingly used as a method for researching career opportunities in hospitality management. Search engines will produce multiple listings of hospitality recruiters with key words such as "hotel manager," "front office manager," and "hotel careers." Also, professional hospitality organizations usually offer a job-posting service on their websites. The Internet provides many opportunities for a new graduate to examine trends that are driving the industry and new technologies that will shape a career in hotel management. This information can assist job applicants in exploring the employment possibilities and prospects in different geographic areas.

Your survey of career possibilities should include a review of a potential employer's economic performance on a balance sheet and other features of its profit-and-loss statement. This information is available on computerized business databases. Take the time to research the economic potential of the company you are considering. Your interview preparations should include reviewing the regional economic prospects and the company's economic performance. This preparation could set the stage for an investment that lasts many years, perhaps a lifetime.

The effort you put into preparing for a visit to a career fair is essential for making this a learning and networking opportunity for you. Questions you could ask a representative of a limited-service hotel include, "What are the typical management responsibilities of an assistant general manager in your organization? What types of visitors frequent your hotels during the week and on the weekend? What is the typical size of your hotels?"

Questions you may want to consider asking a representative of a full-service hotel include, "What size staff is employed in your hotel? Do you have any convention hotels in your portfolio? What services do you typically offer in a hotel in your organization?" These types of questions open lines of communication and help you present yourself as a future professional. Students of hotel management should consider international employment opportunities. Current trade journals allow you to review the many job opportunities for hospitality professionals who have prepared themselves through education and work experience. International employment requires managers to know operations *and* to have a desire to learn and work in another culture.

This option can be very exciting and a great addition to your career. Either, front office manager of The Times Extended-Stay Hotel, received a phone call from the home office of The Times Hotel Management Company asking her to participate in a meeting to discuss the new trend in long-term visitors. The home office is thinking of renovating some of the rooms to attract the guest who wants to stay for 5 to 15 days. How would Esther prepare for this meeting?

This introduced the future professional to the hotel industry. It began with a historical review, including founders of the hotel industry—Statler, Hilton, Ritz, Astor, Waldorf, Wilson, J. W. Marriott and J. W. Marriott Jr., Henderson and Moore, and Schultz. It also discussed historical developments that have shaped the products and services offered to guests, management trends, and economic factors. It covered the atrium concept, marketing and operational emphasis, geographic relocation, the emergence of limited-service hotels, the major reorganization of 1987–1988, adoption of total quality management, technological advances, the continued effects of 9/11, and wireless technology in the hotel industry.

It provided an overview of the industry in terms of types of hotels; market orientation/location (residential, commercial, airport, and highway); sales indicators of occupancy, average daily rate, and RevPAR; levels of service (full-service, all-suites, limited-service, and extended-stay); and type of affiliation or non-affiliation (franchise, referral, company-owned, management contract, and independent ownership). Branding was discussed as an approach to understanding the industry configuration. A review of trends fostering growth in the hotel industry was presented (leisure time, me/pleasure concept, discretionary income, family size, household size, business travel, female business traveller, and travel as an experience). Factors affecting a student's career development choice were discussed, including educational preparation, work experience, professional memberships, ports of entry in a hotel, and researching growth areas in the hospitality industry.

Professor Catherine Vicente has allotted time in the HRI-201 Introduction to Front Office Management course for a field trip this semester. After the first few lectures, she wants to take her class to the hotel establishments in the vicinity of City College. The area is well known for its tourist attractions and is the headquarters of several major U.S. businesses. She appoints a group of students to assist her in setting up tours.

One of the students, Maria, is a resident of the area and suggests they visit the grand old St. Thomas Hotel in the downtown area. She would also like to see a hotel located at the Wide World Airport. Ryan, another student, has worked at a limited-service property in his hometown.

He understands another hotel in that chain is located on the outskirts of the city. David, who is applying for a job at a local hotel, wants to get information on all-suites hotels. Linda has heard of a new extended-stay hotel in town and wants to know what makes it different from a limited-service hotel.

The group has sifted through all the requests and decided to form five teams to visit these places. Each team will appoint one spokesperson for a panel discussion. The spokesperson will present a five-minute summary of what was learned from the visit. What items do you think each spokesperson will include in his or her summary? A recent survey in a suburban community projects an influx of new citizens into the area. Several computer industries will be relocating to the area, and they are expected to employ 25,000 persons at all levels of the organizations. Also, one of these computer companies will locate its corporate headquarters here, with an additional 500 executives arriving soon. The local hotel association has contacted Professor Catherine Vicente of the HRI programme at City College for assistance in determining the impact these new residents will have on their hotels with regard to occupancy and use of facilities. If you were Professor Vicente, what actions would you undertake? Justify your responses with regard to hotel operations and development. If you lived in this community, how would these developments affect your career in the hotel industry?

FOUNDERS OF THE HOTEL INDUSTRY

A history of the founders of the hotel industry provides an opportunity to reflect on our heritage. Learning about the founding giants such as Statler, Hilton, Marriott, Wilson, and Schultz, to name a few, allows a student of the industry to discover the interesting lineage of hoteliers. Studying the efforts of the innovators who carved out the modern hotel industry may help future professionals with their own career planning.

To begin to understand the history of the modern hotel industry, let's look at its early entrepreneurs, who were motivated by wealth and fame on a grand scale. Ellsworth M. Statler (1863–1928) developed the chain of hotels that were known as Statlers, beginning with a hotel in Buffalo, New York, built for the 1901 Pan-American Exposition. Eventually there were Statler hotels in Boston, Cleveland, Detroit, New York City, St. Louis, and other locations. In 1954, he sold the Statler chain to Conrad Hilton. Statler devised a scheme to open an incredible two-story, rectangular wood structure that would contain 2,084 rooms and accommodate 5,000 guests. It was to be a temporary structure, covered with a thin layer of plaster to make it appear substantial, although simple to tear down after the fair closed.

Conrad Hilton (1887–1979) became a successful hotelier after World War I, when he purchased several properties in Texas during its oil boom. In 1919, he bought the Mobley Hotel in Cisco, Texas. In 1925, he built the Hilton Hotel in Dallas, Texas. His acquisitions during and after World War II included the 3,000-room Stevens Hotel (now the Chicago Hilton) and the Palmer House in Chicago and the Plaza and Waldorf=Astoria in New York City. In 1946, he formed the Hilton Hotels Corporation, and in 1948, he formed the Hilton International Company, which came to number more than 125 hotels. With the purchase of the Statler chain in 1954, Hilton created the first major chain of modern American hotels—that is, a group of hotels that follow standard operating procedures in marketing, reservations, quality of service, food and beverage operations, housekeeping, and accounting. Hilton Hotels now includes Hilton Garden Inns, Doubletree, Embassy Suites, Hampton Inns, Harrison Conference Centres, Homewood Suites by Hilton, Red Lion Hotels and Inns, and Conrad International.

Cesar Ritz was a hotelier at the Grand National Hotel in Lucerne, Switzerland. Because of his management abilities, "the hotel became one of the most popular in Europe and Cesar Ritz became one of the most respected hoteliers in Europe." In 1893, William Waldorf Astor launched the 13-story Waldorf Hotel at Fifth Avenue near Thirty-fourth Street in New York City. The Waldorf was the embodiment of Astor's vision of a New York hostelry that would appeal to his wealthy friends by combining the opulence of a European mansion with the warmth and homey qualities of a private residence. Four years later, the Waldorf was joined by the 17-story Astoria Hotel, erected on an adjacent site by William Waldorf Astor's cousin, John Jacob Astor IV. The cousins built a corridor that connected the two hotels, which became known by a single hyphenated name, the Waldorf-Astoria. In 1929, after decades of hosting distinguished visitors from around the world, the Waldorf-Astoria closed its doors to make room for the Empire State Building. The 2,200-room, 42-floor Waldorf=Astoria Hotel was rebuilt on its current site at

Park and Lexington avenues between Forty-ninth and Fiftieth streets. Upon the hotel's opening, President Herbert Hoover delivered a message of congratulations. President Hoover later became a permanent resident of the Waldorf Towers, the luxurious "hotel within a hotel" that occupies the twenty-eighth through the forty-second floors. Conrad N. Hilton purchased the hotel in 1949 and then the land it stands on in 1977. In 1988, the hotel underwent a $150 million restoration. It was designated a New York City landmark in January 1993.

Kemmons Wilson started the Holiday Inn chain in the early 1950s, opening the first in Memphis, Tennessee. He wanted to build a chain of hotels for the traveling family and later expanded his marketing plan to include business travellers. His accomplishments in real estate development coupled with his hotel management skills proved a highly successful combination.

Wilson blazed a formidable path, innovating with amenities and high-rise architecture, including a successful round building concept featuring surprisingly functional pie-shaped rooms. Wilson also introduced the in-house Holidex central reservation system, which set the industry standard for both the volume of business it produced and the important byproduct data it generated (allowing it, for example, to determine feasibility for new locations with cunning accuracy). Wilson died in February 2003 at the age of 90. His legacy to the lodging industry will be serving the traveling public with comfortable, safe accommodations while making a profit for investors.

J. W. Marriott (1900–1985) founded his hotel empire in 1957 with the Twin Bridges Marriott Motor Hotel in Virginia, near Washington, D.C. Marriott Hotels and Resorts had grown to include Courtyard by Marriott and American Resorts Group at the time of J. W. Marriott's death in 1985, at which time J. W. Marriott Jr. acquired the Howard Johnson Company; he sold the hotels to Prime Motor Inns but retained 350 restaurants and 68 turnpike units. In 1987, Marriott completed expansion of its Worldwide Reservation Centre in Omaha, Nebraska, making it the largest single-site reservations operation in U.S. hotel history.

Also in 1987, Marriott acquired the Residence Inn Company, an all-suite hotel chain targeted at extended-stay travellers. With the introduction of limited-service hotels—hotels built with guest room accommodations and limited food service and meeting space—Marriott entered the economy lodging segment, opening the first Fairfield Inn in Atlanta, Georgia, in 1987.

Ernest Henderson and Robert Moore started the Sheraton chain in 1937, when they acquired their first hotel, the Stonehaven, in Springfield, Massachusetts. Within two years, they purchased three hotels in Boston and soon expanded their holdings to include properties from Maine to Florida. At the end of its first decade, Sheraton was the first hotel chain to be listed on the New York Stock Exchange. In 1968, Sheraton was acquired by ITT Corporation as a wholly owned subsidiary, and ambitious development plans were put into place to create a truly global network of properties. In the 1980s, under the leadership of John Kapioltas, Sheraton's chairman, president, and chief executive officer, the company received international recognition as an industry innovator in offering modern hotel accommodations. The Sheraton chain is currently owned by Starwood Hotels and Resorts Worldwide.

In the early 1980s, Ray Schultz founded the Hampton Inn hotels, a company in the Holiday Inn Corporation. These hotels were tagged as limited-service, meeting the needs of cost-conscious business travellers and pleasure travellers alike. Schultz's pioneering efforts in developing a product and service for these market segments have proved a substantial contribution to the history of the hotel industry. At a 1998 celebration of the expansion of the Hampton Inn hotel corporation to more than 800 properties, Schultz said: We started the Hampton Inn chain in 1884 to provide guests with a quality room and special amenities like a free continental breakfast and free local phone calls, all at an exceptional value. The opening of this hotel today tells us that we understood our guests' needs and that price/value is still a viable concept nearly 15 years later. Hampton Inn has been, and is committed to remaining, the standard against which all midpriced, limited-service hotel brands are measured.

The history of the hotel industry is filled with concepts that shaped the products and services offered. The atrium concept design, limited-service hotels, and technology were notable innovations. Management concepts such as marketing and total quality management (TQM) offered managers a new way to do business in hotels. The major U.S. economic reorganization in the late 1980s shaped the way hotels could be profitable. Also, in the 1990s, a new financial approach—real estate investment trusts (REITs)—changed the financial structuring and operation of hotels. The terrorist events of September 11, 2001, continue to affect how hotels market their products and services and deliver hospitality.

The hotel industry has seen many notable developments over the past years. The atrium concept, an architectural design in which guest rooms overlook the lobby from the first floor to the roof, was first used in the 1960s by Hyatt Hotels. The dramatic approach to hotel style [was] exemplified by the Hyatt Regency in Atlanta. Designed by architect John Portman, with a striking and impressive atrium soaring up its 21 stories, the hotel literally changed the course of upscale hotel design. As a result hotels became more than a place to rest one's head. They became hubs for excitement, fun, relaxation and entertainment. This 1,260-room hotel "is now one of the nation's premier convention and trade show facilities with 180,000 square feet of ballroom, exhibit, meeting and hospitality space."

The movement of hotel construction from the downtown, centre-city area to the suburbs in the 1950s coincided with the development of the U.S. highway system. The limited service concept—hotels built with guest room accommodations and limited food service and meeting space—became prominent in the early 1980s, when many of the major chains adopted this way to serve business travellers and travellers on a limited budget. Hampton Inn revolutionized the hotel industry with the creation of the first national brand targeted to the new limited-service hotel segment. The hotels featured spacious, comfortable rooms but eliminated or reduced other elements common to hotels at that time, such as restaurants, lounges, and meeting and lobby space, passing the resulting cost savings on to guests in the form of lower rates. The company pioneered a number of ideas, including the mounting of the first site by a hotel brand on the Internet. In 1989, Hampton Inn became the first national brand to offer guests an unconditional 100 per cent satisfaction guarantee, which today is the cornerstone of all Promus brands and a testament to the company's commitment to quality.

DELIVERY OF ROOM SERVICE MANAGEMENT

ORGANIZE DELIVERY ROOM CARE

This toolkit is designed to provide guidelines for the resuscitation of all infants following delivery using the best information available to date. We believe that the smallest and most immature of infants have unique requirements to ensure an effective transition from fetal to extra-uterine life. These infants have immature organ systems, and without appropriate preparation and intervention can develop severe degrees of hypothermia, and respiratory failure that can significantly increase mortality and morbidity. At the other end of the spectrum, many full-term infants will experience difficulties following delivery, and the basic principles of resuscitation can be applied in a similar fashion for all newborns infants for all gestational ages.

Previous resuscitation guidelines (American Heart Association 2005), while having sections discussing specific issues of prematurity, have not been designed for the most immature of infants. For these infants resuscitation interventions are required more frequently with an almost 100 fold increase in the need for compressions or epinephrine when compared with the term infant. The new guidelines, as reflected in the 6th Edition of NRP and discussed by Kattwinkel et al and Perlman et al reflecting the 2010 International Consensus Conference on Cardiopulmonary Resuscitation and Emergency Cardiovascular Care Science with Treatment Recommendations, have expanded sections on the resuscitation requirements and interventions for the preterm infant. These areas include the recommended use of oximeters and a targeted oxygen strategy to gradually increase the Sp02 for intrauterine values to acceptable neonatal values over at least the first 10 minutes of life using the most appropriate FiO_2; this requires the use of air and an oxygen blender.

While there remains a dearth of prospective research on actual resuscitation interventions for VLBW infants there is some evidence regarding the optimal resuscitation environment and a number of effective practices that can assist in stabilizing VLBW infants. A guiding principle of this toolkit is that whenever possible/feasible, the delivery room/resuscitation room environment should mimic as closely as possible that of the Neonatal Intensive Care Unit, and that every effort should be made to establish homeostasis as soon as possible after birth for all newborn infants.

UTILIZE A STANDARDIZED

The resuscitation and initial stabilization of newborn infants is a transition that consists of several discrete processes and requires the coordination of both personnel and equipment. The methods of quality and process improvement can be used to measure and optimize these processes. We recommend that a standardized, scripted, multidisciplinary approach be utilized to enhance the coordination and guidance of the effort.

Rationale

The resuscitation of all newborn infants requires that a skilled and capable team be present and prepared to offer the most appropriate resuscitation, support and evaluation. However, depending on the urgency of the clinical situation and the relative instability of the newborn, the situation can become anxiety provoking, highly intensive, and even chaotic. Given the multiplicity of the tasks involved, the confined space and limited equipment immediately available, and the variability of the team composition, there are numerous opportunities for miscommunication and even errors in performance and decision-making. The presence of two or more caregivers requires deliberate coordination and optimal communication. In reviewing the processes, personnel and equipment involved in the resuscitation and admission process, we will also present markers of process quality as well as relevant questions that institutions may want to address on a site-specific basis. Finally we will also describe how scripting and rehearsal of resuscitations can lead to improved uniformity of practices and improvement of process measures. The resuscitation of all newborn infants is a well-defined analyzable series of processes that routinely occurs in a specific location, usually conducted by a well-defined team of individuals using appropriate equipment. Process analyses and improvement techniques can be utilized to improve team function in this critical situation.

Preparatory and Initial Management Tasks

Pre-admission activities can be viewed as a series of multiple processes or tasks that may happen simultaneously or in a sequence. Many of these tasks are common to resuscitations performed for newborns at any gestational age, though VLBW infants will need special attention to their respiratory and thermal stabilization. Other high-risk conditions, *e.g.* abdominal wall defects, may necessitate preparation for and application of other interventions and processes.

Therefore, those responsible for resuscitation may want to include education around, and evaluation of, these critical tasks. It is important to appreciate that some of these processes ideally will occur before the resuscitation team arrives in the delivery room. There is a need for discussion with parents on possible limitations to be placed on resuscitative efforts for ELBW infants before or near the edge of viability. While the basics of resuscitation are well covered in the 2011 Textbook of Neonatal Resuscitation Programme the transition from birth to the NICU for VLBWs is more complex and completed only after admission to the NICU.

The following are pre-delivery tasks for any high risk delivery:

Pre-Delivery Tasks

- Review maternal records
- Counsel parents on the need to consider maternal or neonatal transport, if applicable, for more specialized care, addressing both benefits and risks.
- Review relevant morbidity and mortality statistics
- Explain delivery room practices
- Introduce NICU care
- Offer NICU tour
- Discuss breastfeeding, plans for breast milk expression, collection and storage.

The following delivery room and admission tasks are for all newborn infants, unless specified as for VLBW infants:

- Pre-warm radiant warmer
- Prepare polyethylene wrap and hat for VLBW infant
- Check resuscitation bag, oxygen flow and blender, manometer & select an appropriately sized mask, set oxygen blender to preferred concentration for expected infant

- Check suction – bulb and wall
- Check availability of resuscitation medications and other equipment
- Receive baby (sterilely, if need be)
- Dry baby; wrap baby if VLBW
- Evaluate breathing
- Assess pulse and prepare to continuously monitor
- Apply temperature probe and hat
- Apply pulse oximeter
- Respiratory support to the degree indicated
- Initiate support as indicated (*e.g.* none, continuous positive airway pressure, bag/mask, intubation)
- Confirm intubation (auscultation and CO_2 detector)
- Provide ongoing support (controlling peak inspiratory pressure with use of a manometer or a T-piece resuscitator device)
- Assign APGAR scores
- Document
- Prebriefing

Delivery Room Tasks - As Needed

- Cardiac support
- Compressions
- Draw up, administer epinephrine (dose varies by route)
- Volume support
- Obtain emergency vascular access (typically umbilical venous access)
- Draw up, give volume (caution rate of administration)
- Respiratory support
- Draw up and administer surfactant, if done in the delivery room setting
- Obtain cord gases (if not already addressed by Labour and Delivery protocols)

Admission Tasks - As Needed

- Transport to NICU/Stabilization area (to/from resuscitation area)
- Weigh
- ECG monitoring
- Initialize machine, attach leads
- Oxygen saturation monitoring
- Obtain and document initial vital signs
- Establish vascular access
- Place appropriate vascular access device (Umbilical catheter(s), PIV,)
- Provide respiratory support, as indicated
- Set up ventilator, CPAP, hood or nasal cannula device(s) and oxygen as indicated

- Prepare and administer appropriate medications, *e.g.* aquamephyton, and erythromycin eye ointment
- Prepare and administer surfactant, if indicated and not already administered in DR
- Assess gestational age and size
- Screen infant per protocol (*e.g.* blood glucose screening for preterm, SGA, LGA, etc
- Prepare and administer other medications, *e.g.* antibiotics, as indicated
- Obtain initial radiographic, laboratory, and blood gas studies as indicated
- Document information
- For hospitals planning to transport:
- Refer for transport as soon as feasible (even before delivery).
- Care during the First Hour in the non-NICU setting (prior to transport)
- Contact receiving centre as soon as need for transport is identified (prior to delivery or immediately following)
- Provide significant data and condition report
- Obtain care recommendations
- Carry out recommendations, evaluate response and communicate with receiving centre
- Obtain consent for transport and provide family with anticipatory guidance
- Prepare needed documentation for transport
- Copy of maternal and neonatal chart
- Blood sample (maternal and neonatal)
- Duplicate radiology films, laboratory results
- Complete referring hospital section of Neonatal Transport Form
- The STABLE Programme

Implementation Strategies:

The following critical tasks need to be successfully addressed when implementing the standardized, scripted approach:

- Obtain consensus on the policies/procedures that will be utilized for high-risk deliveries
- Develop policy/procedure for special deliveries and determine how it will be communicated that special actions will be needed for a particular delivery
- Ensure that there are systems in place that enable all providers to know the roles for which they have responsibility
- Ensure that there is an evaluation process and that there is feedback about how the actions taken during a specific delivery actually work
- Ensure that there are process or outcome measures to assess, monitor and trend task performance

- Determine how your standardized approach can be adopted and applied in non-standard delivery locations (*e.g.* in the Emergency Department, in an elevator during transport, etc.).

Personnel/Team Composition Optimizing

Accordingly, consideration should be given to understanding the need to identify the key players involved, and their anticipated roles, in each step of the process.

Furthermore, an appreciation of the number of individuals involved is needed to optimize the patterns of communication that constitute effective teamwork. Specific issues regarding personnel, team composition and role assignment include:

Pre-Delivery Counselling

- Ensure that antenatal counselling team is available as necessary
- Team members may include neonatologists, neonatal nurse practitioners, NICU RNs, social workers as available and others appropriate to the situation
- Ensure that obstetrical colleagues can easily identify and contact the counselling team
- Ensure that team members provide all relevant data about morbidity and mortality, as well as the NICU experience
- Data (local and national) about morbidity and mortality of VLBW infants; some centers provide handouts for parents, *e.g.* describing limitations of care at extreme viability.
- Information for parents about NICU care – consents, handouts, video
- Ensures family/medical decisions regarding type and extent of resuscitation are available to the resuscitation team

Team Composition - Delivery Room

- Composition varies widely depending on clinical resources as well as individual patient needs. Team composition should be standardized based on the unique circumstances within each facility.
- Team members must have appropriate training and experience in resuscitation practices & communication.
- At least two health care providers must be available and committed solely for the evaluation and care of the newborn.
- Most teams have a lead, as well as an additional RN or RCP.
- Many hospitals add a RCP or RN as the third member for the highest risk, ELBW infants.
- Team members should be known and immediately available to the delivery room, whether by phone, beeper, or overhead page.
- Labour and delivery room staff should have policies/procedures

addressing the need to contact neonatal resuscitation team in a timely fashion and to provide resuscitation until the team arrives.

- Contingency plans should be created and clarified for multiple deliveries, be they multiple gestation or simultaneous deliveries by multiple mothers.

Team Composition -Admission to NICU

- NICUs or Labour/Delivery may have additional staff, beyond those individuals who attend a delivery, available to assist with the admission tasks.
- Some pre-admission tasks, *e.g.* pre-heating a warmer or obtaining surfactant, preparing intravenous fluids or setting up respiratory support (*e.g.* NCPAP or a ventilator) for a high risk delivery can be addressed by these additional staff members.

Implementation Strategies:

The following critical implementation strategies need to be addressed when implementing the standardized, scripted approach:

- Identify key players for the tasks necessary leading to a newborn stabilization area.
- Identify specific, identifiable patient populations that require a different team composition.
- Insure that team members are readily available and easy to contact.
- Develop contingency plans for forming additional teams to be called to multiple deliveries.

EQUIPMENT AND MATERIALS ISSUES

Delivery Room

- Temperature support – radiant warmer, warm blankets, chemical blankets, caps, polyethylene wrap, temperature probe with insulating cover, and ambient room temperature
- Respiratory support – masks, bags, endotracheal tubes, intubation equipment, CO2 detectors, tidal volume measurement devices, suction
- Monitoring equipment – oxygen saturation monitor & probes
- Crash carts – must accommodate the needs of both maternal and neonatal emergencies.

Equipment Issues -Admission to NICU

- Transport equipment from delivery room to Newborn stabilization area need depends on distance, must include battery power, ventilator, heating source, extra warmed blankets
- Scale
- Radiant warmer

- Monitoring equipment – ECG, O_2 saturation, and cardio-respiratory monitors, and hemodynamic pressure monitor & cables
- Respiratory care equipment -ventilator, CPAP, hood, nasal cannulae, blended oxygen
- Vascular access catheter(s), insertion supplies, & infusion pumps - PIV's, umbilical lines.
- Intravascular fluids -dextrose solutions, TPN, lipids, arterial line solutions.

Implementation Strategies:

The following critical implementation strategies need to be successfully addressed when implementing the standardized, scripted approach:

- Identify what equipment should be available for all deliveries.
- Identify additional equipment needs for specific high-risk deliveries.
- Identify where this equipment can be stored yet remain easily accessible.
- Identify who is in charge of getting equipment to the resuscitation setting in a timely fashion.
- Identify who is responsible for periodic checking of equipment and restocking supplies after use.

QUALITY AND PROCESS IMPROVEMENT METHODOLOGIES

Analysing and understanding the processes, personnel and equipment essential to the resuscitation of all newborn infants is necessary in order to improve the reliability and quality of those resuscitations. The following is a list of quality improvement methodologies that may be particularly applicable to neonatal resuscitation.

Quality Improvement Methodologies

- Scripting/Role Modelling
- Process Mapping -fishbone diagrams, value stream mapping
- Lean Thinking (reducing wastes of time, materials)
- Evaluating Process Parameters – track and trend them over time.

Given the multi-factorial nature of long-term outcomes of VLBW infants, it is unlikely that an institution could ever prove that implemented changes in resuscitation directly improved such outcomes. Therefore, we must utilize short-term care process and clinical measurements as indicators of the quality of our resuscitative efforts.

This approach is supported by the fact that some of these process measurements, such as admission temperature, have been associated with long-term outcomes of VLBW infants. Other measurements will reflect the preparedness of the team and the consistency of the resuscitative efforts, both of which should be associated with improved clinical outcomes.

Resuscitation Process and Quality Measurements

- Measurements of personnel and equipment readiness; evaluations of resuscitations and track and trend them over time.
- Time to surfactant administration
- Admission temperature
- Initial ABG -pO_2, pCO_2 values in acceptable range
- Time to completion of an established stage of stabilization (*i.e.*, respiratory support (NCPAP/ETT, etc) in place, oximeter functioning, lines in place, blood glucose determined using a point of care testing, weight, vital signs (including BP), temperature, time of temperature, laboratory specimens sent). If one defines a set of agreed upon initial stabilization steps, looking at the timing of reaching this stage of NICU admission may be a very helpful metric for improving care. While such a standard is not reported in CPQCC databases, a single centre could conceivable track this over time as a quality metric.

Planning and scripting care

Given the analysis described above, the question arises as to how to plan and practice for the provision of the highest quality resuscitation of newborn infants as well as for their admission to the NICU. Uniformity of practice is more likely to lead to the following care improvements:

Improved consistency in performing resuscitation tasks.

- eases teaching of practices to new participants
- facilitates identification of outlying practices and process-associated outcomes
- increases ease in evaluating and comparing alternative practices with one another

Evidence of improvement associated with consistency in practice is found in data from non-medical fields, including high-risk activities such as commercial aviation, nuclear energy and the military. The benefit of such a planned, scripted approach in health care can be found in reports on managing cardio-vascular surgeries (Edmondson 2001) emergencies and trauma. For neonatal resuscitations, the benefits are as follows:

Planning/Scripting:

- Defines separate roles with common goals
- Clearly outlines tasks of each team member (may be defined each shift, may use identifying flash cards, may want to include recorder)
- Facilitates awareness and communication between team members
- Establishes a timeline
- Allows for rehearsal

Cooperation & Communication:

- Involve multiple team members in establishing goals

- Enables viewing of protocols by all team members
- Facilitates review and revision of the script by the multi-disciplinary team in order to establish and amend the time line and role assignments
- Encourages team learning of resuscitation strategy, incorporating "super-users"
- Enhances communication
- Allows for continuous, safe real time feedback
- Educates staff as to the rationale behind interventions

Facilitating uniformity:

- Allows widespread visibility of the script
- Encourages use of the plan of care present at the bedside
- Supports use of orders including ventilator orders
- Other times of standardization could include identify complex delivery types and hospital guidelines: Hydrops, Abdominal wall or spine defect, etc

HOTEL STAFF MANAGEMENT

BACKGROUND AND SIGNIFICANCE

This is a pilot study designed to help in the development of a large-scale research project on the future of work in the hospitality industry. The hospitality industry has experienced a wave of restructuring, consolidation, and new practices to cut costs including lean staffing and greater performance demands on the workforce.

The study will explore the impact of these and related changes on the health and quality of life the largest occupational group within the hospitality industry, room cleaners. Jobs in housekeeping (and food and beverage) operations of the hospitality industry represent the future of work in this growing service sector with jobs characterized by increasing repetitive physical workloads, low income, low skill utilization, low job control, and virtually no prospects for training and career advancement.

There is compelling evidence that this kind of low-income jobs result in a disproportionate high burden of illness, injury, and disability. Unlike other services, businesses in the hospitality industry, cannot be moved across the borders but will remain a fast growing sector in industrialized countries, creating increasing societal costs including disproportionate workers' compensation and health care utilization costs. Working conditions in the hospitality industry are representative for a growing number of workers of colour, working women, and former welfare recipients. The hospitality industry has become a major target for welfare-to-work and job training programmes in cities throughout the country. An aging workforce is faced with increasing job demands in a competitive industry constantly upgrading services. Few studies have been done on this growing workforce and little is known about their specific working conditions and health risks and how they could be addressed.

This study will take a broad view at physical and psychosocial job factors, health, including general self-rated health, musculoskeletal disorders, work-related injury, and health-related quality of life. Further, this study will investigate which innovative organizational strategies can help to reduce the burden of illness and disability in the service sector by comparing hotels in different market segments, different degrees of unionization, and hotels with innovative joint labour management health and safety programmes, negotiated child care/elder care benefits, and a multi-employer worker training project.

San Francisco is expected to provide several union and non-union cases for comparison to similar cases in low union density markets and to a selected number of additional cases in other high union density markets that highlight organizational innovations (such as training and career laddering in Las Vegas or work redesign in New York to avoid sub-contracting). The pilot study focuses on housekeeping since it is the largest occupational category in the industry and is the central focus of a larger study including hotels in other cities. The study combines qualitative and quantitative research methods and is participatory in design and implementation thereby laying the groundwork for sustainable efforts in understanding and improving health and safety of service workers.

COOPERATIVE SETTING

The pilot study is administered by the director of the Labour and Occupational Health Programme (LOHP) at UC Berkeley Robin Baker, in collaboration with Niklas Krause, physician epidemiologist from the School of Public Health at UC Berkeley. Funding is provided by the United Way of the Bay Area and Local 2 of the Hotel Employees and Restaurant Employees International Union (HERE).

The level of union density in San Francisco is among the highest in the country. The largest union is HERE Local 2 which represents 8,000 workers or around 70 % of the industry's non-supervisory workforce. In San Francisco the multi-employer bargaining agreement with about 50 hotels in the city has taken wages and benefits out of competition.

The starting wage for non-tipped jobs is $11 per hour plus comprehensive portable benefits. San Francisco presents a special opportunity for comparisons of work conditions in the industry and for future workplace interventions because of a unique partnership with twelve first class hotels that is designed to improve customer service and job quality in the industry.

The San Francisco Hotels Partnership Project has implemented a core-training programme throughout much of the industry, and developed pilot projects to resolve conflicts at the lowest possible level within an organization. The parties have also worked to establish joint decision making processes for redesigning work in the food and beverages and housekeeping operations.

OBJECTIVES

The objectives of the proposed study are:

- To identify key changes in technology and organization in the hospitality industry that affects the future of work for room cleaners.
- To explore the impact of industry changes on the health, injury rates, and well being of room cleaners.
- To develop potential strategies for enhancing the future health and well being of room cleaners.
- To demonstrate the utility of a participatory research methodology that actively involves room cleaners in every aspect of the study, and that results in the ability of participants to effectively advocate for health-promoting changes in their work.
- To explore the relevance of the findings to similar low wage populations, particularly in the service sector.
- To widely disseminate study findings and recommendations to encourage further study and action by academic researchers, labour, management, policy makers, and advocacy groups.

The pilot consists of establishing an advisory group of participants, focus groups with room cleaners, the development and administration of a survey instrument, analysis and discussion of results with participants, and dissemination of results in a final report.

Although designed for comparisons with other cities, the pilot will allow for comparisons between high quality and low quality (in terms of labour relations and customer service) hotels in San Francisco. In addition, the use of some standardized instruments will enable comparisons of working conditions of housecleaners with jobs in different industries. Finally, future studies are planned to investigate the effects of changes in work conditions and labour-management collaboration on injury incidence, health and functional/quality of life outcomes.

Progress

A 30-member advisory group of room cleaners participated in the planning and implementation of this pilot study. Ten focus groups have been conducted with about 20-25 housecleaners at each meeting recruited from different San Francisco hotels. A survey instrument has been developed and tested in a focus group of 25 roomcleaners. The revised survey has been translated into Spanish and Chinese languages. The survey includes standard instruments such as Karasek's Job Content Questionnaire, Siegrist's effort/reward balance scales including measures of intrinsic and extrinsic effort, SF36 health-related quality of life/functioning, general health status, and specifically developed questions to measure physical workload, musculoskeletal symptoms, work injury, and demographic variables. Depending on reading and language skills it takes 30 to 75 minutes for workers to complete the survey with assistance available from researchers and translators. 139 out of 397 roomcleaners from 4 selected hotels completed the survey during the first wave of data gathering in December 1998.

Plan

A second wave of data gathering among 258 non-responders is planned for February. Intensive efforts of personal recruitment at the workplace and telephone follow-up is planned for a random 50% sample of 130 non-responders.

Data management and basic analyses limited to descriptive statistics and comparisons between four hotels are planned for March and April. A draft report will be prepared for the union by May 15 and will be available for joint labour management focus groups preparing for upcoming bargaining of five-year contracts. Further analyses including comparisons with national survey data, discussions with participants, and a final report will be completed by December 1999. Publications in peer reviewed scientific journals may be prepared simultaneously or thereafter but no funds have been allocated yet for this purpose.

Specific Aims

- Establish cooperative relationships between researchers and stakeholders
- Select 2-6 hotels (with a total of about 400 employees) representing "good" and "bad" overall work conditions as perceived by union representatives and key senior employees
- Conduct a series of focus groups with room cleaners from different hotels which are not selected into the pilot survey, to (a) gain a detailed understanding of their job tasks, workload, working conditions, and work-related health problems, (b) to discuss the scope and goals of the research project, and (c) to invite further participation in design and execution of the study (*e.g.* development and administration of survey instrument, organization of meetings with roomcleaners at participating hotels)
- Develop draft survey instrument
- Discuss draft survey instrument with union representatives and other key informants
- Pre-test survey instrument in focus group with 25 roomcleaners
- Revise survey instrument
- Translate survey instrument into Spanish and Chinese languages

- Obtain names and addresses of current employees at selected hotels
- Invite employees of selected hotels by mail to fill out the survey in off-site meetings
- Conduct first wave of survey under supervision of University researchers and with assistance of translators at different times and locations in San Francisco
- Create database for data entry using Microsoft Access software
- Enter data from questionnaires and check for completeness
- Follow-up of workers with incomplete responses by telephone in their first language
- Update list of employees and draw random 50% sample from survey non-responders
- Obtain contact information including telephone numbers for non-responders
- Invite all non-responders by mail to fill out the survey in off-site meetings
- Invite 50% random sample of non-responders by telephone and/or personal contact to fill out the survey in off-site meetings
- Conduct second wave of survey
- Provide follow-up meetings with research team for workers willing to participate but unable to attend the pre-scheduled survey meetings
- Enter data into database
- Check data for missing information and inconsistencies
- Prepare final data set with variable labels for all data fields and create scales for work conditions based on theoretical and/or empirical approaches
- Analyze survey data and (a) prepare frequency tables for all questionnaire answers and scales stratified by hotels, (b) compare responses of first wave responders with second wave of responders in 50% sample intensively followed up, (c) compare responses between "good" and "bad" hotels, (d) compare work conditions with reports from studies in other industries, (e) investigate the association between work conditions and self-reported general health, musculoskeletal symptoms, quality of life, and work injury
- Gather OSHA 200 logs in selected hotels for 1998 and enter in separate database
- Replace employee names by ID code in injury data set and merge with survey data
- Analyze merged injury (OSHA 200 log) and survey data, and, specifically (a) compare frequency of self-reported and formally reported work injuries, (b) examine self-reported reasons for not reporting a work-related injury, (c) investigate associations between

working conditions and both self-reported and OSHA 200 log based injury frequency

- Prepared draft report on May 15, 1999
- Present report to stakeholders in joint labour management meetings
- Prepare final report by December 1999
- Prepare peer-reviewed journal article(s)
- Integrate results of pilot study into proposed larger study comparing working conditions and health outcomes across different US cities representing different organizational settings and labour markets in the hotel industry.

PRELUDE TO STAFFING

Staffing is the third sequential function of management. Up until now the executive housekeeper has been concerned with planning and organizing the housekeeping department for the impending opening and operations.

Now the executive housekeeper must think about hiring employees within sufficient time to ensure that three of the activities of staffing—selection (including interviewing), orientation, and training—may be completed before opening. Staffing will be a major task of the last two weeks before opening.

The development of the Area Responsibility Plan and the House Breakout Plan before opening led to preparation of the

Department Staffing Guide, which will be a major tool in determining the need for employees in various categories. The housekeeping manager and laundry manager should now be on board and assisting in the development of various job descriptions. The hotel human resources department would also have been preparing for the hiring event.

They would have advertised a mass hiring for all categories of personnel to begin on a certain date about two weeks before opening.

Even though this chapter reflects a continuation of the executive housekeeper's planning for opening operations, the techniques described apply to any ongoing operation, except that the magnitude of selection, orientation, and training activities will not be as intense. Also, the fourth activity—development of existing employees —is normally missing in opening operations but is highly visible in ongoing operations.

Job Specifications

Job specifications should be written as job descriptions are prepared. Job specifications are simple statements of what the various incumbents to positions will be expected to do. An example of a job specification for a section housekeeper is as follows:

Job Specification—Example

Section Housekeeper (hotels) [often Guestroom Attendant —GRA] The incumbent will work as a member of a housekeeping team, cleaning and servicing for occupancy of approximately 18 hotel guestrooms each day. Work will generally include the tasks of bed making, vacuuming, dusting, and bathroom cleaning.

Incumbent will also be expected to maintain equipment provided for work and load housekeeper's cart before the end of each day's operation. Section housekeepers must be willing to work their share of weekends and be dependable in coming to work each day scheduled. [Any special qualifications, such as ability to speak a foreign language, might also be listed.]

Employee Requisition

Once job specifications have been developed for every position, employee requisitions are prepared for first hirings (and for any follow-up needs for the human resources department). Note the designation as to whether the requisition is for a new or a replacement position and the number of employees required for a specific requisition number. The human resources department will advertise, take applications, and screen to fill each requisition by number until all positions are filled.

For example, the first requisition for GRAs may be for 20 GRAs. The human resources department will continue to advertise for, take applications, and screen employees for the housekeeping department and will provide candidates for interview by department managers until 20 GRAs are hired. Should any be hired and require replacing, a new employee requisition will be required.

Bibliography

A.S. Bisht: *Conservation of Textiles, Bone and Ivory*, Agam Kala Prakashan, Delhi, 2010.

Ajeet Jaiswal: *Anthropo-Medical Profile of Textile Workers*, Alfa Publication, Delhi, 2012.

Anil K. Kanungo *China and WTO : Issues and Challenges in Textiles and Telecommunications*, Shipra Publication, Delhi, 2015.

Anita Tyagi: *Digital Textile Designing*, Sonali Publication, Delhi, 2011.

Charles O'Neill: *Cosmo Dictionary of Dyeing and Calico Printing : Containing A Brief Account of All the Substances and Processes in Use in the Arts of Dyeing and Printing Textile Fabrics with Practical Receipts and Scientific Information*, Cosmo Publication, Delhi, 2008.

D. Gopalakrishnan and T. Karthik: *Basics of Textile Chemical Processing*, Daya Publication House, Delhi, 2016.

Dick Kooiman: *Bombay Textile Labour: Managers, Trade Unionists and Officials 1918-1939*, Manohar Publication, Delhi, 2000.

Ela Dedhia and M. Hundekar:*Ajrakh : Impressions and Expressions : A Journey of Antique Traditional jndian Textile "Printing with Natural Dyes" from Its Origin Till Date*, Colour Publication, Delhi, 2008.

Fredrick W Bunce: *Buddhist Textiles of Laos, Lan Na and the Isan : The Iconography of Design Elements*, D K Printworld Publication, Delhi, 2003.

Giles Tillotson: *A Passionate Eye : Textiles, Paintings and Sculptures from the Bharany Collections*, Marg Publications, Delhi, 2014.

H. Panda: *A Concise Guide on Textile Dyes, Pigments and Dye Intermediates with Textile Printing Technology*, NIIR Project Consultancy Services , 2013.

J. M. Matthews: *Applications of Dyestuffs : To Textiles, Paper, Leather and Other Materials (2 Vols-Set)*, Abhishek Publication, Delhi, 2014.

K.D. Saksena: *Dynamics Of Indias Textile Economy : Towards A Pragmatic Textile Policy*, Shipra Publication, Delhi, 2002.

M.K. Khandelwal and M.L. Chauhan: *Dyeing Printing and Textile*, Ritu Publication, Delhi, 2004.

Parul Bhatnagar: *Decorative Design History in Indian Textiles and Costumes*, Abhishek Publication, Delhi, 2005.

Rajesh Kalra: *Bleaching of Textile Fibres*, Abhishek Publication, Delhi, 2011.

Rosemary Crill: *Chintz Indian Textiles for the West*, Mapin Publication, Delhi, 2008.

Shantha Hariharan: *Cotton Textile and Corporate Buyers in Cottonpolis: A Study of Purchase and Prices in Gujarat 1600-1800*, Manak Publication, Delhi, 2002.

Sudhir Gupta and Ankur Gupta: *Complete Technology of NONWOVENS : Fabrics, Carry Bags, Composites, Geotextiles, Medical Textiles, Fibres, Felts, Apparels, Spunlace and Absorbent Nonwoven*, EIRI Publication, Delhi, 2013.

Sushama Swarup: *Costumes and Textiles of Awadh from the Era of Nawabs to Modern Times*, Roli Books Publication, Delhi, 2012.

Swati Bhargava: *Clothing and Textiles : (A Practical Aspect)*, Shree Niwas Publication, Delhi, 2010.

William H. Dooley: *A Compendium of Textiles*, Abhishek Publication, Delhi, 2011.

Index